Volcanoes

Volcanoes

FOURTH EDITION

Robert Decker
and
Barbara Decker

W. H. Freeman and Company · New York

Publisher: Susan Finnemore Brennan
Acquisitions Editor: Valerie Raymond
Development Editor: Sharon Merritt
Marketing Manager: Scott Guile
Media/Editor: Victoria Anderson
Project Editor: Bradley Umbaugh
Cover and Text Designer: Diana Blume
Illustrations: Fine Line Illustrations
Illustration Coordinator: Bill Page
Photo Editor: Patricia Marx
Production Coordinator: Julia DeRosa
Composition: Matrix
Manufacturing: RR Donnelley

Library of Congress Control Number: 2005929602

ISBN: 0-7167-8929-9 (EAN: 9780716789291)

Printed in the United States of America

First printing

W. H. Freeman and Company
41 Madison Avenue
New York, NY 10010
Houndmills, Basingstoke, RG21 6XS, England

www.whfreeman.com

Contents

About the Authors

Until his death, two months before the publication of this book, Robert Decker was active in the field of volcanology. He was Scientist Emeritus with the U.S. Geological Survey, having spent 5 years as Scientist-in-Charge at the Hawaiian Volcano Observatory. He was on the faculty of Dartmouth College for 25 years. Barbara Decker is a science writer specializing in natural history and national parks. The husband-and-wife team has twelve other books to their credit, including *Mountains of Fire: The Nature of Volcanoes* (Cambridge University Press, 1991), *Volcano Watching* (Hawaii Natural History Association, 1997), and *Volcanoes in America's National Parks* (Odyssey Guides, 2001). Between 1986 and 2005, they have written eight road guides to national parks, the most recent being *Road Guide to Lassen Volcanic National Park* (Double Decker Press, 1997).

Preface

Volcanoes assail the senses. They are beautiful in repose and awesome in eruption; they hiss and roar; they smell of brimstone. Their heat warms, their fires consume; they are the homes of gods and goddesses. Volcanoes are described in words and pictures, but they must be experienced to be known. Their roots reach deep inside the Earth; their products are scattered in the sky. Understanding volcanoes is an unconquered challenge. This book poses more questions than answers; such is the harvest of curiosity.

Any book about volcanoes is bound to be highly descriptive, but we have tried to look behind these spectacular phenomena and to emphasize the processes involved. "How do volcanoes work?" is the key question.

In the 24 years since the first edition of *Volcanoes* was published, there has been a great increase in public interest in and awareness of volcanism. More than 250 volcanoes have erupted in that time, most more than once; 30,000 people have been killed by those eruptions. That the most lethal eruptions during this period were from volcanoes not on our original list of "The World's 101 Most Notorious Volcanoes" is a sure sign that volcanology is still an evolving science.

But there has been good news during these years, too. Scientists have had some notable successes in forecasting explosive eruptions, particularly in the case of Mount Pinatubo in the Philippines in 1991. Although more people have continued to crowd into cities and rich farmlands in the shadow of potentially dangerous volcanoes, during this same time scientists have gained a better understanding of volcanoes, and governing authorities have gained more respect for their potential danger. In the near future there will be close, life-or-death races between population pressure and successful eruption forecasting, with unknown outcomes. Most important, new ideas, new discoveries, and worldwide cooperation have helped geologists add more pieces to the puzzle of how volcanoes work.

The fourth edition of *Volcanoes* incorporates significant new material:

- Both Mount Pinatubo and Yellowstone National Park are given full chapter coverage (new chapters 5 and 12), underscoring their importance in our understanding of volcanic systems and life cycles.
- Coverage of Mount St. Helens includes the 2004 eruption.
- The book now concludes with a third new chapter, "Volcanoes in the Solar System," putting Earth's volcanoes into galactic perspective.
- In the past few years the World Wide Web has become a virtual world library available to anyone who has a computer and an online connection. To capitalize on this important new source of information, we have supplemented our book with many colored illustrations available on the W. H. Freeman and Company Web site (www.whfreeman.com/volcanoes4e) and provided Web site links at the end of each chapter where more information related to that chapter can be found.

Acknowledgments

The list of sources at the end of each chapter reflects only part of our debt to the hundreds of students of volcanoes whose general knowledge of volcanic processes and products we have used. Three friends deserve our special thanks: James G. Moore of the U.S. Geological Survey, the late Allan Cox of Stanford University, and the late John Staples of W. H. Freeman and Company. Without their help and encouragement, the first edition of this book would not have been written. Two important general sources of information about volcanoes have become available since the previous edition: *Encyclopedia of Volcanoes*, edited by Haraldur Sigurdsson (Academic Press, 2000), and *Volcanism*, by Hans-Ulrich Schmincke (Springer, 2004).

During the evolution of four editions of this book we have been greatly helped by many people. We gratefully acknowledge the following reviewers and editors:

Reviewers: Andy Barth, Indiana University; Mark Brinkley, Mississippi State University; Elwood Brooks, California State University; Allan Cox, Stanford University; John Eichelberger, University of Alaska; Albert Eggers, University of Puget Sound; Karen Grove, San Francisco State University; Michael W. Hamburger, Indiana University; Paul Jewell, University of Utah; Ian Lange, University of Montana; Otto Muller, Alfred University; H. R. Naslund, State University of New York at Binghamton; William Rose, Michigan Technological University; Jim Taranik, University of Nevada; David West, Middlebury College; Craig White, Boise State University.

W. H. Freeman and Company personnel: Diana Blume, Susan Finnemore Brennan, Scott Guile, Patricia Marx, Sharon Merritt, Valerie Raymond, Norma Roche, and Bradley Umbaugh.

Our thanks to all of you,
Robert and Barbara Decker

Introduction

Most of the Earth's mountains, valleys, and plains have been slowly sculpted by uplift and erosion over the last million years, and these, in turn, have formed from other landscapes long vanished. Volcanoes, though, operate on a different time scale.

On the morning of May 18, 1980, Mount St. Helens in Washington State was a tall, conical volcano 2950 meters high; by nightfall it was an ugly stump of a mountain 2550 meters high, with a cloud of gas and ash rising from a new horseshoe-shaped crater (Plates 13 and 14). A massive avalanche had torn away the whole north side of the mountain, unleashing a giant steam explosion that shattered the mountaintop, devastated a forest of 10 million trees, and killed 57 people. A huge eruption cloud jetted upward from the volcano, blanketing central Washington with ash and moving as far as the East Coast in only three days.

Volcanologists from all over the world converged on Washington State in a concerted effort to find answers to the specific questions of just what happened and why, as well as the larger questions of how the lessons learned at Mount St. Helens could be applied to other dangerous volcanoes. Both applied and basic research benefited. Volcano-monitoring tools and techniques were refined, and the assessment of volcanic hazards in other potentially active areas was greatly improved.

The explosion of Mount St. Helens caught the rapt attention of the public too, when Americans who had thought of a volcanic eruption as something that happened in a Flintstones cartoon suddenly found volcanic ash sifting down into their backyards. They, too, wanted answers to such questions as, why are volcanoes like Mount St. Helens explosive while others, like those in Hawaii, pour out quieter streams and fountains of lava? Why do volcanoes occur in chains like the Cascade Range or the Japanese islands? How does one know when a volcano is dead and not just dormant?

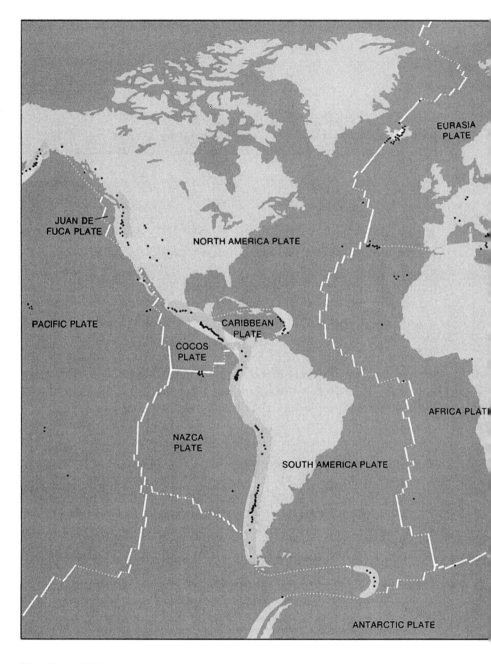

The Ring of Fire. (After Tom Simkin, J. D. Unger, R. I. Tilling, P. R. Vogt, and H. Spall, *This Dynamic Planet: World Map of Volcanoes, Earthquakes, Impact Craters, and Plate Tectonics*, U.S. Geological Survey, 1994.)

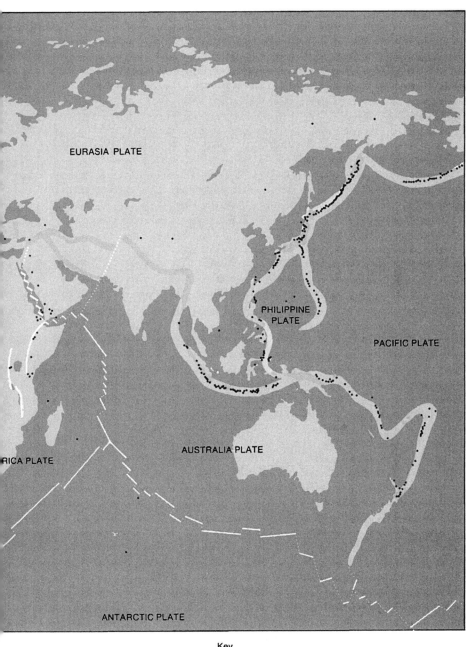

EURASIA PLATE

PHILIPPINE
PLATE

PACIFIC PLATE

AUSTRALIA PLATE

RICA PLATE

ANTARCTIC PLATE

Key

| Rift zones | Subduction zones |
| Strike-slip (transform) faults | · Geologically young volcanoes |

Mount St. Helens was still in the public consciousness when, five years later, another volcanic catastrophe shocked the world. In November 1985, Nevado del Ruiz Volcano in Colombia erupted, triggering mudflows that raced down its steep valley, engulfing a whole town and killing more than 20,000 people.

In the aftermath of this tragedy, scientists focused their attention on studying not just the details of the actual eruption but also the problem of reducing volcanic risk by improved monitoring of hazardous volcanoes and by the use of more effective techniques for forecasting eruptions. Also of major concern was, and still is, the sensitive social and political problem of how and when to issue a forecast of a disaster that may or may not take place. Scientists do not want to be accused of crying "wolf"; the dilemma is that sometimes the wolf really is there, but may not be hungry.

Forecasting of the eruption of Mount Pinatubo in the Philippines in 1991 was more successful. Although this giant explosive eruption—the second largest of the twentieth century—killed 300 people, 80,000 were evacuated from the most dangerous areas in the few days before the cataclysm. Fine volcanic ash and sulfuric acid aerosol droplets injected into the stratosphere by the Pinatubo eruption circulated worldwide and had a measurable effect on global climate.

In 2004, Mount St. Helens surprised geologists with a new eruption. With sophisticated scientific equipment placed in safe and strategic positions and full media coverage, it not only offered a valuable learning opportunity but also put on quite a show.

Mount St. Helens, Nevado del Ruiz, and Pinatubo are all located on the Ring of Fire—the belt of young mountains, earthquakes, and volcanoes that encircles the Pacific Ocean. Why are volcanoes not randomly scattered about the Earth? We start with this question of why volcanoes are common in some regions of the Earth and not in others.

CHAPTER 1

Seams of the Earth

1   San Andreas Fault cutting across the Elkhorn and Carrizo plains in south central California. The North America Plate is on the right, and the Pacific Plate is on the left. Movement on the fault is called right lateral movement because for a person standing on either plate, the sense of motion on the opposite plate is to the right. (Photograph by Robert E. Wallace, U.S. Geological Survey.)

The matching shores of eastern South America and western Africa form an intriguing jigsaw puzzle. Did Brazil's bulge once fit against the Congo? Did some great supercontinent break up and drift apart, each piece becoming one of our present continents?

Alfred Wegener, an Austrian scientist, championed the concept of continental drift for 20 years until his death on the Greenland Ice Cap in 1930. He noted not only that the edges of the continents make a rough fit, but also that the Appalachian mountain range in eastern North America breaks off abruptly in Newfoundland and reappears across the Atlantic Ocean in Ireland, Scotland, and Scandinavia. He argued, too, that the similarity of ancient European and American fossils and the dissimilarity of more recent fossils was one more piece of evidence that the continents as we know them are still-drifting segments of an ancient supercontinent.

Wegener also recognized that the Earth's surface has two predominant elevations: a continental one between sea level and 1000 meters above sea level, and an ocean basin depth between 4000 and 5000 meters below sea level (Figure 2). He hypothesized that this difference in elevation between the continents and the oceans was caused by differences in the density and thickness of the rocks in the continental blocks compared with that of the oceanic crust. He believed that the lighter and thicker continental blocks floated above the oceanic crust like icebergs drifting through the sea.

To understand Wegener's use of the terms *float* and *drift*, one must understand the great span of geologic time. Wegener envisioned the oceanic crust as being more viscous than tar or pitch, so that the "floating" took thousands of years to reach a balance, and the "drifting" proceeded even more slowly. Wegener's idea of continental drift was one of the great scientific controversies of its day, and his arguments in favor of it were alluring. Wegener called the great supercontinent he envisioned Pangaea: it was so large that it contained both the polar scars of ancient ice ages and the equatorial legacy of vast coal swamps that existed before the continent's breakup. According to his theory, the present locations

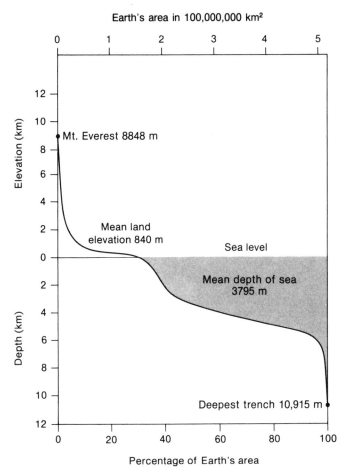

2 The Earth's surface elevations and surface areas. The two benches on the curve reflect the basically different thicknesses and densities of continental and oceanic crust. (Adapted from Alyn C. Duxbury, *The Earth and Its Oceans.* © Addison-Wesley, Reading, Massachusetts. Fig. 3.2. Reprinted with permission.)

of glacial deposits near the equator and of coal deposits near the current poles were the result of continental drift.

But Wegener was also a zealot, and many of his contemporaries — wary of too much zeal — considered him a crackpot. In the first quarter of the twentieth century, geophysicists who specialized in the study of the Earth's interior were especially skeptical because Wegener was unable to identify the strong forces needed to propel the continents away from one another. Because these experts had already concluded that the interior of the Earth was as solid and strong as steel, the notion that floating continents could drift through such a strong seafloor was untenable.

These geophysicists' knowledge of the deep interior of the Earth came mainly from analyzing what happens to earthquake waves as they propagate through the Earth, a branch of earth science called seismology. Its methods are similar to the doctor's technique of thumping on your chest to find out whether there is liquid or air in your lungs: the quality of the "thump" sound passing through your chest provides the key. Seismology is a rigorous discipline based on precise physical principles, and its conclusions are well respected by all geologists. The denial of continental drift because of such geophysical objections was convincing despite the good circumstantial evidence in its favor, and the hypothesis went underground for 30 years.

Among geologists, however, there is a half-joke called Smith's law, which states that anything that did happen, can happen. So even without understanding how the large horizontal movement of continents could happen, some geologists still looked for evidence that it did happen.

After World War II, seafloor topography was mapped by continuous echo sounders on oceanographic ships. In addition, seismologists interested in the locations of earthquakes plotted patterns of epicenters, which coincided with the crests of the submarine ridges mapped by the oceanographers. Without openly declaring themselves, many earth scientists began to think that maybe Wegener was right.

In the fall of 1963, Harry Hess, a professor of geology at Princeton University, gave his presidential address to the Geological Society of America, which was entitled "Further Comments on the History of Ocean Basins." It was a bombshell. In his lecture, Hess built on a proposal made in 1928 by Arthur Holmes, an eminent British geologist, that the seafloor may be spreading apart (Figure 3). In Hess's elaboration, molten rock erupts from the Earth's interior along mid-ocean ridges, creating new seafloor that slowly spreads away from the ridges. To prevent this spreading from expanding the Earth, Hess suggested, the seafloor eventually sinks into the deep oceanic trenches. With Hess's address, Wegener's largely discredited ideas gained new legitimacy, and the earth science revolution began.

Evidence that helped prove Wegener and Hess right came in the early 1960s from marine geophysicists who were analyzing the magnetic field of the ocean floor. They found that the magnetic field was unusually strong directly over the mid-ocean ridges. The reason, they proposed, was that the rocks making up the ridges contained iron-rich minerals and were magnetized in parallel with the Earth's magnetic field, which tends to reinforce that field.

As the geophysicists extended their surveys away from the ridges, they found that the magnetic field at the surface of the sea formed a zebra-

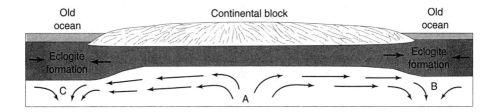

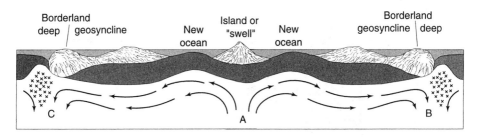

3   Original 1928 drawing by Arthur Holmes illustrating continental drift and the development of new ocean basins. Eclogite is a dense rock formed under high pressure that has the same general composition as basalt. In Holmes's own words, "A current ascending at A spreads out laterally, extends the continental block and drags the two main parts aside, provided that the obstruction of the old ocean floor can be overcome. This is accomplished by the formation of eclogite at B and C, where sub-continental currents meet sub-oceanic currents and turn downwards. Being heavy, the eclogite is carried down, so making room for the continents to advance." (From *Transactions of the Geological Society of Glasgow*, vol. 18, 1928–1929, p. 579. Reprinted in A. Holmes, *Principles of Physical Geology*, 2nd ed. New York, Ronald Press, 1965, p. 1001.)

striped pattern of alternating belts of unusually high and unusually low intensity. Their explanation of the low-intensity belts was that the ocean floor beneath them had formed at times when the Earth's magnetic field pointed to the south, rather than to the north as it does today.

By the time of the marine magnetic surveys in the early 1960s, geophysicists studying the magnetic properties and ages of lavas on land had already demonstrated that the Earth's magnetic field periodically switches from north to south and back again. The time between reversals may be as short as 30,000 years or as long as several million years. Using this magnetic reversal time scale, the marine geophysicists knew when the field had flipped, and so they could estimate the age of the ocean floor beneath each magnetic stripe (Figure 4). The stage was set for a dramatic new idea.

Geologists now proposed that instead of the continents drifting through the ocean floor, the seafloor spread away from the oceanic ridges as it formed, carrying along the continents as part of large, spreading

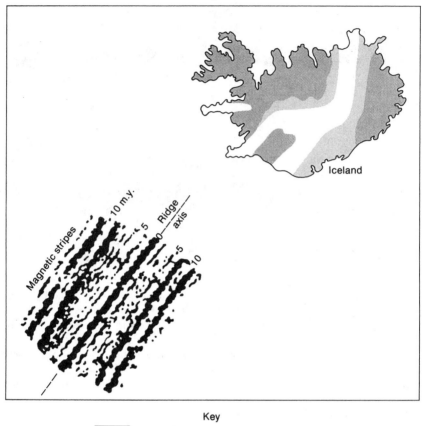

Key

☐ Youngest rocks, less than 10,000 years old

▨ Rocks 10,000 to 4,000,000 years old

▧ Rocks 4,000,000 to 15,000,000 years old

m.y.   Millions of years

4 Magnetic field strength measured at the sea surface southwest of Iceland shows a definite striped pattern parallel to the axis of the Mid-Atlantic Ridge. The symmetry of the pattern on either side of the ridge and the similar pattern in the ages of volcanic rocks in Iceland are strong evidence that the seafloor spreads away from the ridge axis. (Adapted from Charles Drake, "The Geological Revolution," Condon Lectures, Oregon State System of Higher Education, Eugene, OR.)

plates. This new idea was called *seafloor spreading* or *plate tectonics*. The magnetic stripes provided a record of both the geometry and the rate of spreading. The pattern and ages of the magnetic stripes parallel to the ocean's ridge systems thus provided a giant slow-motion tape recording of the Earth's horizontal plate movements over the past 170 million years.

The pattern that has emerged shows the Earth's surface broken into ten large and five smaller plates that are moving in various directions relative to one another (Figure 5). The margins or seams between the plates are of three basic types: *convergent*, in which one plate overrides the other; *divergent*, in which the plates separate and move apart; and *strike-slip* or *transform*, in which the plates slide by each other without separating or overriding (Figure 6). Convergent and divergent margins are characterized by different types of major volcanic activity, but transform margins have much less or no volcanism, so the link between plate tectonics and volcanoes is close indeed.

Earthquakes, topography, and other geologic structures identify the three types of seams. At the convergent margins, also known as *subduction zones*, one plate (generally the oceanic one) slips beneath the other. This movement causes earthquakes, which originate either near the surface seam or in the sinking plate at depths as great as 700 kilometers (Figure 7). Physical forces at the convergent margin cause the ocean floor to bend down into a deep trench, while the edge of the upper plate is pushed into folds or is broken by *thrust faults* (in which the upper side moves up and over the lower side). Rock layers scraped off the ocean floor are stacked into slices or folded against the upper plate. The geologic result is an island arc like Japan or a mountain chain like the Andes.

At divergent margins, the separating plates often form a valley, or *rift zone*. The earthquakes are shallow and follow the center of the rift. No folds are formed because the area is being stretched, not squeezed, and the fractures are either open cracks or slumps called *normal faults*. Most divergent margins are submarine, located along the axes of the Earth's mid-oceanic ridges. New oceanic crust is formed at the separating edges and spreads outward from the ridges. The main ridges are the Mid-Atlantic Ridge, the East Pacific Rise, and the Indian Ocean ridges. These ridges are broad, rugged, submarine mountain ranges a few kilometers high, a few thousand kilometers wide, and tens of thousands of kilometers long. Drain away the oceans, and you would be able to see the greatest mountain system on Earth.

The transform margins are also mainly submarine. They connect offsets of the oceanic ridges into a rectangular pattern that, on topographic maps of the seafloor, looks like an alligator's back (Figure 8).

Sometimes a slice of continent gets entangled in the shearing of a transform margin. The San Andreas Fault in California is a classic example of a transform fault on land. Los Angeles is moving northwest rel-

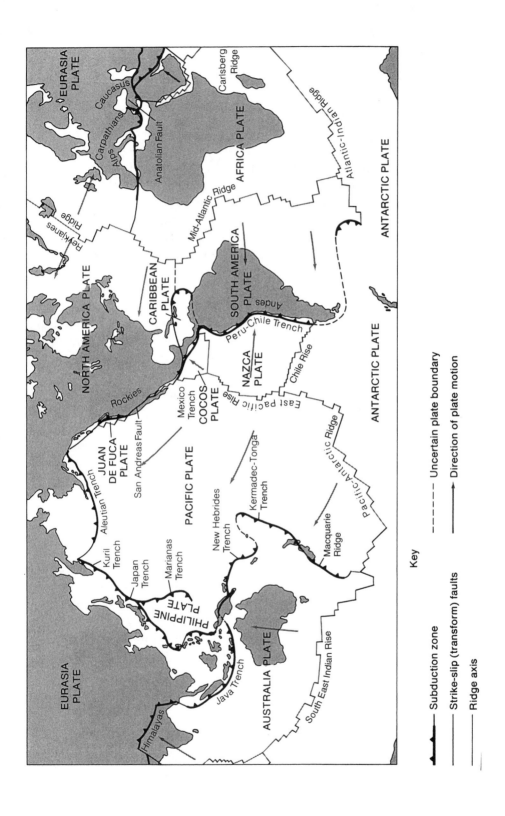

Key

Subduction zone — — — Uncertain plate boundary

Strike-slip (transform) faults ⟶ Direction of plate motion

Ridge axis

Plates and features labelled:

EURASIA PLATE

Caucasus

Carpathians

Alps

Anatolian Fault

Carlsberg Ridge

AFRICA PLATE

Atlantic-Indian Ridge

ANTARCTIC PLATE

Reykjanes Ridge

Mid-Atlantic Ridge

NORTH AMERICA PLATE

CARIBBEAN PLATE

SOUTH AMERICA PLATE

Andes

Peru-Chile Trench

NAZCA PLATE

Chile Rise

Rockies

Mexico Trench

COCOS PLATE

East Pacific Rise

ANTARCTIC PLATE

JUAN DE FUCA PLATE

San Andreas Fault

Aleutian Trench

PACIFIC PLATE

Pacific-Antarctic Ridge

Kermadec-Tonga Trench

Macquarie Ridge

Kuril Trench

Japan Trench

Marianas Trench

New Hebrides Trench

PHILIPPINE PLATE

EURASIA PLATE

Java Trench

Himalayas

AUSTRALIA PLATE

South East Indian Rise

5 Rigid plates of the Earth's surface are slowly moving horizontally away from and toward each other. The arrows shown assume that the Africa Plate is not moving. Plates separate along the crests of mid-ocean ridges, slide past each other along strike-slip faults, and converge at subduction zones. (Adapted from J. F. Dewey, "Plate Tectonics." Copyright © 1972 by Scientific American, Inc. All rights reserved.)

ative to San Francisco at a rate of about 5 centimeters per year. Even so, society will probably make them one big city long before nature slides the two together. Shallow earthquakes, linear valleys, and twisted rocks mark these transform margins (see Figure 1).

Despite the overwhelming evidence that seafloor spreading takes place, the strength and solidity of the Earth's interior remained a problem for the idea of continental drift. The problem was finally resolved by seismologists, which is ironic, since it was the seismologists whose arguments had shot down Wegener's hypothesis in the 1920s. When they

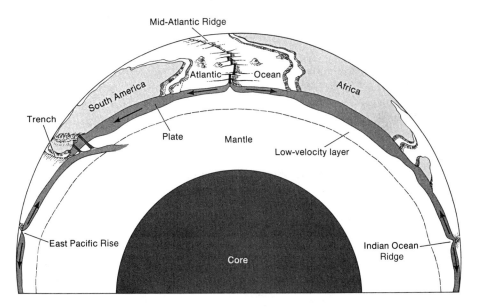

6 Movement of plates shown in a cross section of the Earth. The Africa and South America plates separate along the Mid-Atlantic Ridge at a rate of a few inches per year. The South America and Nazca (eastern Pacific Ocean) plates converge to form the Andes mountain range. The thickness of the plates and of the low-velocity layer on which they move are exaggerated so that they can be shown at this small scale. (Adapted from K. C. Burke and J. Tuzo Wilson, "Hot Spots on the Earth's Surface." Copyright © 1976 by Scientific American, Inc. All rights reserved.)

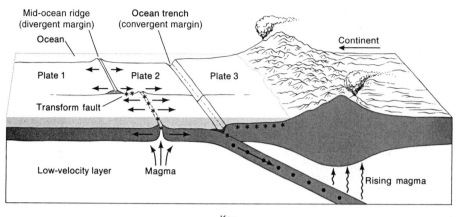

Key

* Shallow earthquakes
(tension on ridges; lateral slip on transform faults)

• Deep earthquakes
(mainly showing thrusting and down-dip compression)

7 A schematic cross section of plate margins, showing the association of earthquakes with the plate boundaries formed by mid-ocean ridges, strike-slip (transform) faults, and subduction zones. (Adapted from L. Sykes et al., in *Earth*, F. Press and R. Siever, p. 644. Copyright © 1974, W. H. Freeman and Company.)

looked closely at earthquake wave records, particularly those from underground nuclear bomb tests with precisely known locations and times of origin, the seismologists discovered a zone about 100 to 200 kilometers beneath the Earth's surface where seismic waves travel unusually slowly and are partially absorbed.

This low-velocity layer is not very strong because it probably contains a small percentage of molten rock. In the nomenclature of plate tectonics, this weak layer is called the *asthenosphere,* and the strong plate above it the *lithosphere.* The plates move on the plastic asthenosphere under the influence of small horizontal forces, and fractures in the Earth's lithosphere at the divergent plate margins allow the molten rock to leak to the surface, forming volcanoes (Figure 9).

8 A topographic map of the Atlantic Ocean with the water drained away. The Mid-Atlantic Ridge is one of the world's greatest mountain ranges. The breaks perpendicular to the valley crest of the ridge are called *fracture zones* or *transform faults* —other terms for strike-slip margins. (From *World Ocean Floor Panorama* by Bruce C. Heezen and Marie Tharp, 1977. Reproduced by permission of Marie Tharp, all rights reserved.)

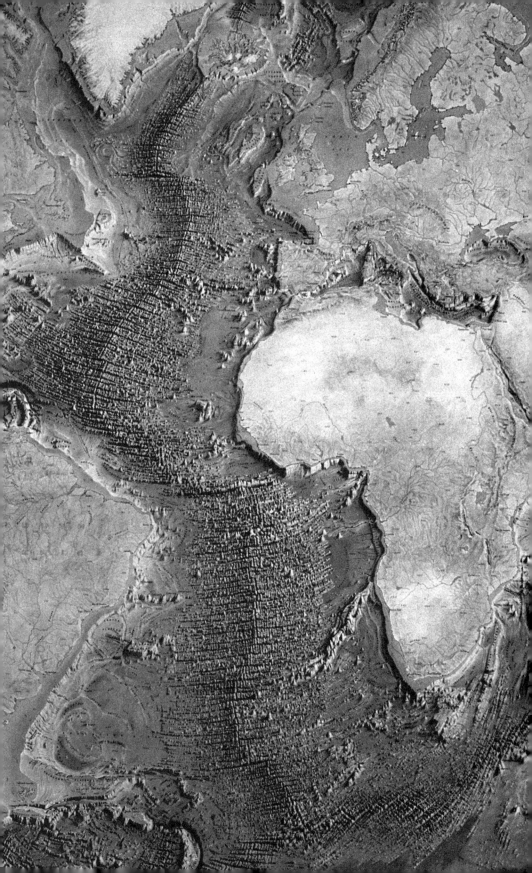

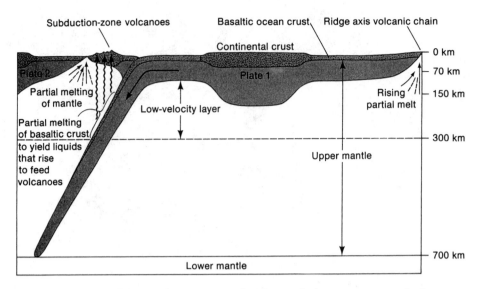

9 A cross section of the Earth's upper mantle. The rigid plates are composed of solidified rock that moves on the partly molten, low-velocity layer. The plates are approximately 70 kilometers thick under the oceans and 100 to 150 kilometers thick under the continents. The continents are part of the plates and so move with them. (Adapted from J. F. Dewey, "Plate Tectonics." Copyright © 1972 by Scientific American, Inc. All rights reserved.)

What are the great forces that slowly move the plates? The options are simple—push, pull, slide, or drag—but the arguments are more complex. All these forces or, more precisely, *stresses* (forces divided by the areas on which they act), have their advocates, whose general arguments are as follows.

Geologists who vote for "push" envision a prying apart of the plates at their divergent margins. They suggest that the injection of magma into the cracks between the plates pushes them apart. Those who vote for "pull" believe that the subducting part of a plate probably is more dense than the rocks in the upper mantle that surround it, which causes it to sink and pull down the rest of the plate behind it. Advocates of this slab-sink concept presently appear to be in the majority. Their argument is that the oceanic plates with the longest subduction zones, such as the Pacific Plate, are moving the most rapidly.

The problem with both push and pull is that they require the plates to be strong enough to be moved by forces acting on their edges. But because the plates are huge slabs of the Earth's crust and upper mantle, their very size makes them inherently weak. Opponents of push or pull use the well-known rule that to make a small-scale model behave realistically, you must reduce its strength in proportion to the change in scale. So the opponents propose a thought experiment: Squeeze out a line of toothpaste on a sink top and then try to move the toothpaste tube by pushing or pulling on the end of the line of paste. The proponents of push or pull counter that if you made the sink top slippery enough and pushed or pulled slowly enough, it might work.

Some geologists argue that gravity moves the plates from locations that are generally higher (the mid-ocean ridge axes) to locations that are generally lower (the oceanic trenches). Gravity works internally on the whole plate—as a so-called body force—avoiding the strength problem posed by forces acting only on or near the plate margins. According to this idea, a plate literally slides downhill from ridge to trench.

Dragging a plate along with the flow of plastic rock in the upper mantle beneath it—the original suggestion of Arthur Holmes—also avoids the strength problem, but does require some frictional connection between the bottom of the plate and the moving mantle material. Whereas gravity sliding requires a very slippery contact zone beneath a plate, dragging by currents slowly stirring in the upper mantle requires a more viscous contact zone.

Possibly the plates are propelled by a combination of stresses, one more dominant in some regions, another in other places. For now, the question of what exact stress moves a particular plate has not, in our opinion, been clearly answered, but unanswered questions are the lifeblood of scientific research.

Wegener turned out to be a prophet, although his original idea has been significantly changed. He believed that the continents drifted like isolated rafts through a viscous oceanic crust. The theory of seafloor spreading envisions rigid plates of oceanic crust forming and spreading from the mid-ocean ridges, with the continental blocks carried like rafts frozen into the larger plates, which are composed of both oceanic and continental crust.

In the last five decades of unprecedented progress in understanding the Earth, the development and rapid acceptance of the concept of seafloor spreading stands as one of the great revolutions in science. Today we can not only reconstruct the jigsaw puzzle of the continents, but also track their ongoing movements. The dynamics of plate motions control both earthquakes and volcanoes, which occur, for the most part, on the creaking and leaking margins of the plates.

The association between volcanoes and earthquake belts has long been recognized. Aristotle proposed that earthquakes were caused by the rumbling of pent-up gases that were eventually vented through volcanoes. He recognized the connection, even though we would not agree today with his explanation. Perhaps the most important lesson of the past 40 years has been the virtue of scientific humility: Who knows which ideas that now appear untenable will turn out to be right, and which of those ideas that we currently accept on the basis of available evidence will seem hopelessly naive 2000 years from now?

It is now recognized that most of the world's volcanoes occur along either the seams of converging plates (the subduction zones) or the seams of diverging plates (the rift zones) (see Figure 9). Hence, plate tectonics plays a major role in the location and nature of volcanic activity. The next five chapters amplify this tectonic framework. Chapter 2 describes the birth of Surtsey, a rift volcano in Iceland, and Chapter 3 explains the general features and processes related to rift volcanoes. Chapters 4 and 5 examine individual volcanoes—Mount St. Helens and Pinatubo—near the converging edges of plates, and Chapter 6 looks at the general features and processes related to subduction volcanoes.

Links

This Dynamic Earth: The Story of Plate Tectonics, U.S. Geological Survey
http://pubs.usgs.gov/publications/text/dynamic.html

This Dynamic Planet Map, U.S. Geological Survey
http://pubs.usgs.gov/pdf/planet.html

National Academy of Sciences, Beyond Discovery: When the Earth Moves: Seafloor Spreading and Plate Tectonics
http://www.beyonddiscovery.org/content/view.article.asp?a=229

World Topography, National Oceanic and Atmospheric Administration
http://www.ngdc.noaa.gov/mgg/image/2minrelief.html

World Earthquakes, U.S. Geological Survey
http://neic.usgs.gov/neis/general/seismicity/world.html

Vectors of Plate Motion, Jet Propulsion Laboratory
http://sideshow.jpl.nasa.gov/mbh/series.html

Plate Tectonic Data, Cascade Volcano Observatory
http://vulcan.wr.usgs.gov/Glossary/PlateTectonics/

Sources

Glen, W. *The Road to Jaramillo.* Stanford, CA: Stanford University Press, 1982.

Kearey, P., and F. J. Vine. *Global Tectonics.* Palo Alto, CA: Blackwell Scientific, 1990.

Kious, W. J., and R. I. Tilling. *This Dynamic Earth: The Story of Plate Tectonics.* Denver: U.S. Geological Survey, 1996.

Moores, E. M., and R. J. Twiss. *Tectonics.* New York: W. H. Freeman, 1995.

Oreskes, N., ed., with Homer Le Grand. *Plate Tectonics: An Insider's History of the Modern Theory of Earth.* Boulder, CO: Westview Press, 2001.

Press, F., R. Siever, J. Grotzinger, and T. H. Jordan. *Understanding Earth*. New York: W. H. Freeman, 2003.

Simkin, T., J. D. Unger, R. I. Tilling, P. R. Vogt, and H. Spall. *This Dynamic Planet: World Map of Volcanoes, Earthquakes, Impact Craters, and Plate Tectonics*. Denver: U.S. Geological Survey, 1994.

Stein, S., and J. Freymueller, eds. *Plate Boundary Zones*. Washington, DC: American Geophysical Union, 2002.

Sullivan, W. *Continents in Motion*. New York: McGraw-Hill, 1991.

Tarbuck, E., and F. Lutgens. *The Theory of Plate Tectonics* (CD-ROM). Taos, NM: Tasa Graphic Arts, 2003.

CHAPTER 2
Surtsey, Iceland

10 Lightning bolts discharge the buildup of static electricity in the uprushing ash column at Surtsey. Ninety-second time exposure. (Photograph by Sigurgeir Jónasson, December 1, 1963.)

Iceland sits astride the Mid-Atlantic Ridge, part of the worldwide rift system. This 300-by-500-kilometer island whose north shore touches the Arctic Circle is entirely volcanic in origin. Eruptions occur about every five years, sometimes from distinct volcanoes like Mount Hekla, which has erupted twenty-three times since A.D. 900, but often from long fissures that erupt only once. Iceland grows in size from volcanic eruptions and is worn away by the storms of the North Atlantic. The three most recent volcanic eruptions occurred at Vatnajokell in 1996, at Grimsvotn in 1998, and at Hekla in 2000. Icelandic farmers and fishermen know that the Earth is a dynamic place; they all are geologists at heart.

The story of the Surtsey eruption begins at sea. The late Icelandic volcanologist Sigurdur Thorarinsson was its chronicler; this account is taken mainly from his works.

At 7:15 A.M. on November 14, 1963, Olafur Vestmann, the cook on a fishing boat off the south coast of Iceland, noticed dark smoke on the horizon against the half-light of dawn. Since there was no land in that direction, his first thought was that it was a ship on fire, but soon the fishermen recognized that the black columns of volcanic ash signaled an eruption from beneath the sea. By 11:00 A.M., the eruption cloud of ash and steam had reached a height of 3500 meters, and volcanologists were on their way to observe the eruption from the air.

At 11:30 A.M., the eruption was coming from three submarine vents along a northeast–southwest line. Explosions from the vents were occurring every few seconds, shooting jets of dense black ash to heights of 100–150 meters. By 3:00 P.M., the separate eruption columns had joined along a 500-meter line across the ocean's surface, and the rapid emission of black volcanic debris indicated that a new island was forming. Surtsey, named for Surtur, the mythological giant of fire, was born that night (Figure 11).

The eruption was a surprise. Only in hindsight were the few clues to the forthcoming eruption recognized. Two days before the eruption, a marine research vessel noticed a strange rise in the temperature of the sea surface above the normal 7°C–9°C in an area about 3 kilometers

11 Column of steam and ash rising more than 3000 meters from the Surtsey eruption on November 16, 1963, two days after the eruption was first sighted. (Photograph by Hjálmar Bárdarson.)

from the eruption site. Also on November 12, people in Vik, a coastal village 80 kilometers to the east, had noticed the rotten-egg smell of hydrogen sulfide. The seismograph at Reykjavík, 120 kilometers away, had recorded weak tremors a week before the eruption was first observed, but the location of the seismic disturbance could not be determined. No preliminary earthquakes were felt at the fishing port of Vestmannaeyjar on Heimaey Island, 22 kilometers northeast of Surtsey, the closest settlement to the eruption site (Figure 12). Presumably the eruption began quietly at 130 meters below sea level and took days or weeks to build the vol-

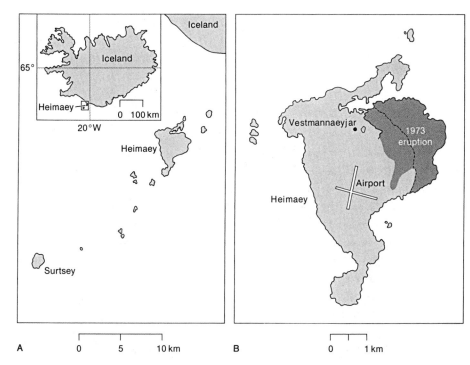

12 A. Map of the Vestmann Islands, showing the location of Surtsey, which formed in 1963–1967, and Heimaey, which formed in prehistoric times. B. The shaded area represents lava added to Heimaey during a 1973 eruption. The former shore is shown by a dashed line.

cano to just beneath the sea surface. There, the explosive activity could no longer be contained and quenched by the pressure and chill of the sea.

During the first week, the new island erupted and grew almost continuously. Closely spaced explosions merging into a steady jetting of volcanic ash formed a towering column that rose to a height of 9 kilometers. By November 18, Surtsey was a ridge 550 meters long and 45 meters high, split lengthwise by the erupting fissure. Gradually one vent along the fissure became dominant, and the island began to assume a more circular shape. On November 24, the island was 900 meters long and 650 meters wide, and the rim of the main crater was almost 100 meters high. The entire island was built of loose, dark volcanic debris, thrown out in fragments by the explosive eruptions. Lava flows of liquid rock had not yet begun, and the loose volcanic pile was an easy target for the strong winter storms of the North Atlantic. But the eruption more than kept pace with the marine erosion and slumping. By February 5, 1964, the maximum height of the volcanic island was 174 meters, and the maximum diameter was 1300 meters, including a wave-cut terrace about 150 meters wide around the shore.

During the first three months of the eruption, the sea had easy access to the erupting vents, both by direct flooding and by seepage through the loose volcanic pile. This contact with water produced the billowing steam explosions that characterize shallow submarine eruptions. Each explosion expelled a black mass of rock fragments, out of which shot numerous larger fragments of pasty lava, called *volcanic bombs*. The flying bombs left trails of black volcanic ash that rapidly turned white and furry as the hot, invisible steam in the trails cooled and condensed. The sight of the black, jet trail–like arcs turning grayish white made the eruption look like an exploding fireworks factory (Figure 13). The bombs whistled as they spun in flight, some landing in the sea more than a kilometer from the vent. The steam explosions themselves were peculiarly quiet, beginning with a muted thump and a nearly silent burst of debris.

13   Black jets of ash explode to heights of 250 meters during the Surtsey eruption. Jet trails follow behind large volcanic bombs that whistle as they spin in flight. (Photograph by Thorleifur Einarsson.)

When the explosions occurred in rapid sequence, as they often did, a large cloud of ash and steam rose above the island. On rare calm days this plume reached as high as 10 kilometers, but generally strong winds bent the column at 500–2000 meters and caused ashfalls many kilometers downwind, contaminating the water supply caught on the roofs of Vestmannaeyjar. The violent updrafts of air into the ash cloud caused whirlwinds, waterspouts, lightning flashes, and sometimes hailstones of falling volcanic fragments coated with ice (see Figure 10).

At times the explosions merged into a continuous uprush of volcanic fragments and steam. Bombs, glowing red to orange, lit the rising column at night. There were only brief lulls in the eruption during November and December, but by January, periods of repose lasted up to a day. As the island grew, the seawater was more effectively blocked from the crater, and the steam explosions diminished.

In April, fire fountains and lava flows became dominant, and formed a cap of hard, solid rock over the lower slopes of Surtsey (Figure 14). The island was now assured a place on the world's maps; the waves still

14   During 1964, Surtsey grew until the sea was blocked from the vent area. As a result, the eruptions changed from explosions to incandescent lava fountains and flows. White steam clouds rising where the flows entered the sea replaced the black and gray explosion clouds. (Photograph by Captain Gardar Pálsson.)

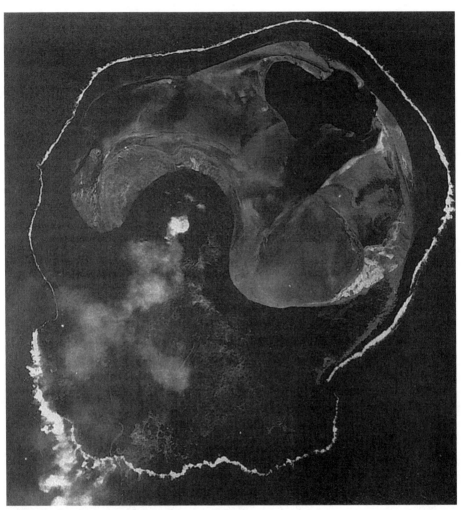

A

15 Aerial photographs looking down vertically (north at top) on the island of Surtsey.
A. The island was about 1.3 kilometers wide, east to west, on August 25, 1964. The gray
area is volcanic ash and cinders, and the dark area in the south is covered by basaltic
lava flows from the crater. A second crater developed to the east of the active vent
between 1964 and 1967.

pounded on the shores, but they met their match on the hardened lavas.
The quiet effusion of lava continued for more than a year as Surtsey increased in area to 2.5 square kilometers, more than half of this capped
by hard lava flows (Figure 15).

During 1965 and 1966, small explosive eruptions built low islands
of volcanic debris both northeast and southwest of Surtsey, but since these

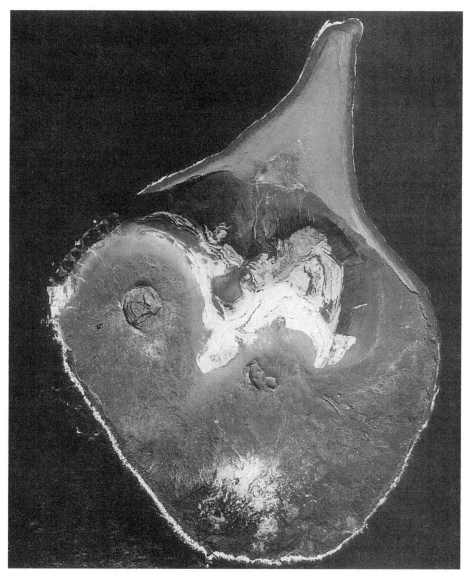

B

15 B. By July 21, 1996, a large area of lava on the south side of the island had been eroded away by the strong waves of the North Atlantic Ocean, and a sand spit had grown on the lee side of Surtsey. (Photographs © by the National Land Survey of Iceland.)

loose volcanic piles were not large or armored with lava flows, they were soon washed away by ocean storms.

In August 1966, lava again erupted from the main crater on Surtsey, and lava flowed continuously in tunnels beneath its hardened crust to the edges of the island. We visited Surtsey during this last activity and were amazed to see that the orange tongues of lava pouring from the tunnels into the sea still glowed as they plunged beneath the waves. The insulating effects of the lava's chilled skin and the steam layer on its surface were so effective that they prevented rapid quenching and steam explosions.

By June 1967, the eruption was over; it had lasted for three and a half years. The total volume of volcanic ash and lava was slightly more than 1 cubic kilometer, only 9 percent of which was above sea level. The rates of eruption were highest in the early phases, diminishing more or less continuously during the course of the volcanic activity. The temperatures of the lava on emission averaged about 1140°C.

The eruption had three main phases. First, there was the quiet, undetected growth of Surtsey from 130 meters below sea level to just a few meters below sea level, possibly lasting for several weeks. Second, explosive eruptions from shallow water built an island of loose volcanic debris, the beginning of the observable eruption. Third, the blocking of water from the vent stopped the explosions and allowed the quiet emission of lava flows (Figure 16).

The shape of Surtsey reflects the island's three-part volcanic history. The sides of the underwater base are steep, owing to the rapid cooling and piling up of the submarine lavas. The explosive debris formed steep crater walls and cones above sea level, but was cut into a flat bench at sea level by wave erosion. The final lava flows formed gentle slopes where they poured out and hardened over the loose debris from earlier explosions.

Surtsey was born in sea, steam, and fire, its rock ribs molded by the environment into which it erupted. The Earth is shaped by such conflicting forces, but over immense spans of time; it is rare to witness these events condensed into less than four years.

In one sense, the story of Surtsey is just beginning. The new island is closed to all but researchers studying the ways in which, even in this hostile environment, life arrives and slowly but tenaciously develops into a complex ecological system. Recently we asked our colleague Páll Einarsson, an Icelandic seismologist and volcanologist, about how life has taken hold on Surtsey since 1967. His answer was "slowly," and he elaborated as follows: Birds began nesting on the island three years after its formation, and the various species now living there all depend on the sea for their food. Some soil has formed in the birds' roosting places, and these have become areas of vegetative cover. Many species of plants and insects now call Surtsey home. Seals have been breeding on the island since 1983, but they are more transients than residents.

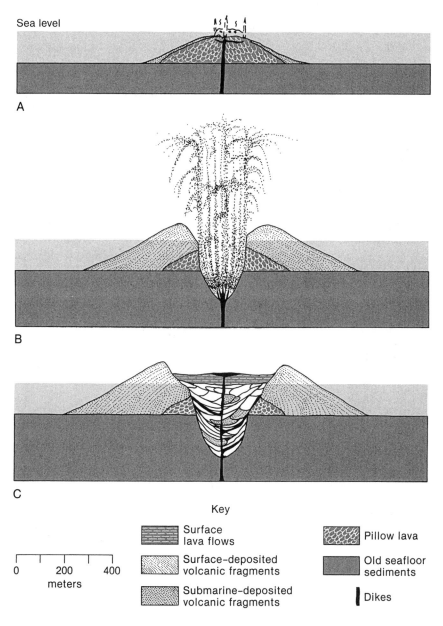

Sea level

A

B

C

Key

Surface lava flows

Surface-deposited volcanic fragments

Submarine-deposited volcanic fragments

Pillow lava

Old seafloor sediments

Dikes

0   200   400
meters

16 Diagrammatic cross sections showing the three main phases of the eruption of Surtsey. A. The submarine volcano of pillow lava is just reaching the sea surface and becoming explosive. (Pillow lava forms during submarine eruptions where the pressure of the overlying water prevents explosive boiling; see Chapter 3.) B. Continuous jetting explosions from the contact of seawater with erupting lava excavate the crater downward into old seafloor sediment. C. The collapse of walls fills the deep crater, and the slumps are covered by surface flows of fluid basaltic lava. These cross sections are schematic profiles drawn along a west northwest–east southeast line through the center of the crater, as seen in Figures 15A and 15B. Data on the deposits that slumped into the deep crater were obtained by drilling. (Adapted from J. G. Moore, *Geological Magazine* 122 [1985]: 655.)

From empty sea, through chaos and barren island, to the slow beginnings of life, the story of Surtsey recapitulates in a small way the origin of the Earth.

Links

Surtsey, Volcano World, 2001
http://volcano.und.nodak.edu/vwdocs/volc_images/europe_west_asia/surtsey.html

Surtsey Research (in Icelandic), Surtsey Research Society, 2004
http://www.ni.is/surtsey/

Surtsey photos, Surtsey Research Society, 2004
http://www.surtsey.is/pp_ens/photo_map_1.htm

U.S. Central Intelligence Agency: The World Factbook: Iceland, 2003
http://www.cia.gov/cia/publications/factbook/geos/ic.html

Iceland information, 2004
http://www.geocities.com/davidjreith/island.html

Surtsey, Jorgen Aabech, 2003
http://www.vulkaner.no/n/surtsey/esurtmenu.html

Surtsey, Explore North, 2004
http://www.explorenorth.com/library/weekly/aa042601a.htm

Nordic Volcanological Institute, 2004
http://www.norvol.hi.is/

Sources

Friðriksson, Sturla. *Surtsey: Evolution of Life on a Volcanic Island.* New York: Wiley, 1975.

Jakobsson, Sveinn, Sturla Friðriksson, and Erlingur Hauksson. *Surtsey 30 ára.* Reykjavík: Surtseyjarfélagid, 1993. (In Icelandic.)

Magnússon, Borgþór, og Sigurður H. Magnússon. "Vegetation succession on Surtsey, Iceland, during 1990–1998 under the influence of breeding gulls." *Surtsey Research* 11 (2000): 9–20.

Thorarinsson, Sigurdur. *Surtsey.* New York: Viking, 1967.

Thorarinsson, Sigurdur. "Surtsey: Island Born of Fire." *National Geographic*, May 1965.

CHAPTER 3

Fire under the Sea

17 Hawaiian lava entering the sea. (Photograph by R. W. Grigg, University of Hawaii.)

In nature things move violently to their place, and calmly in their place.
— Francis Bacon (1561–1626)

What do Surtsey and other rift volcanoes tell us about the formation of the Earth? They tell us that most of the deep ocean floor, 60 percent of the Earth's surface, is of volcanic origin — the product of rift volcanism.

The worldwide rift system is 60,000 kilometers long, and nearly all of it is submarine. Where it does appear on land, in places like Iceland, the volcanic activity is probably greater than it is along the submarine ridges — great enough, in fact, to build the ridge above sea level. The reason for this difference in volcanic activity is a topic of lively debate, and we will return to it in Chapter 8. Most geologists now agree, however, that Iceland is reasonably typical of the volcanic and structural processes that occur on the mid-ocean ridges, and an island is certainly easier to study than an underwater ridge.

Geologists who have studied the mid-ocean ridges believe that reservoirs of molten rock, known as *magma chambers*, exist at depths as little as 1 to 3 kilometers beneath the ridge axes. The height and width of each of these shallow bodies of partly molten rock is only a few kilometers, but in total they are tens of thousands of kilometers long, separated into segments along the crest of the oceanic ridges. Molten rock from these chambers feeds into the spreading rift and temporarily scars over the cracks between the separating plates (Figure 18).

The number of volcanoes that erupt every year to fill these submarine cracks may be greater than that recorded on land. Although a deep submarine volcanic eruption has never been observed, a rough estimate of the number of such events can be made from the observation that Iceland averages one eruption every five years. Iceland represents about $1/200$ of the length of the spreading oceanic ridges. To account for its extra elevation, assume that Iceland's eruption rate is twice that of an equal length of the submarine ridges, whose volcanoes therefore erupt once every ten years. Thus 200 Icelands times $1/10$ of an eruption per year equals 20 eruptions per year along the oceanic ridge system. The fact that not one of these deep submarine eruptions has ever been directly witnessed demonstrates that many frontiers remain to be crossed in our exploration of the Earth. It is only a matter of time, however, until these deep submarine eruptions are seen and studied.

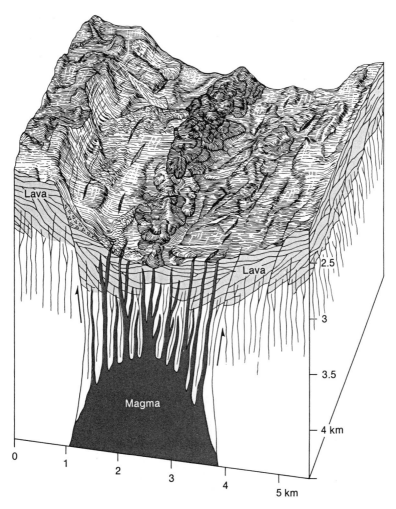

18 A diagram of the structure and submarine topography of the central rift valley of the Mid-Atlantic Ridge. The vertical scale is exaggerated two times. (After R. Hekinian, J. G. Moore, and W. B. Bryan, *Contributions to Mineralogy and Petrology* 58 [1976]: 107.)

There already have been some close encounters. In 2002, submariners reexploring the Galápagos Rift, where submarine hydrothermal vents were first discovered in 1977, saw convincing evidence that areas mapped in 1989 were now covered by fresh basaltic lava flows. On the Juan de Fuca Ridge, off the coast of the northwestern United States, an earthquake swarm and thermal anomaly in June 1993 was interpreted as a possible submarine lava eruption. Dives with submersibles have confirmed the eruption of a new 1.6-million-square-meter flow at that site.

The volume of volcanic rock needed to heal the tear between the spreading plates is enormous. Every year 2.5 square kilometers of new seafloor are formed by the spreading of the plates. The volcanic part of the rocky plates is 1 or 2 kilometers thick at their spreading edges; hence, the volume of new volcanic rock amounts to about 4 cubic kilometers of new oceanic crust every year.

The description of the Surtsey eruption in Chapter 2 contains an interesting paradox. The eruption of molten rock from a vent at shallow depth in the sea led to violent steam explosions. But the lava that flowed through tunnels to the growing edge of Surtsey was seen to flow beneath the waves and to generate white steam clouds without violently disrupting the molten rock. Both were fire under the sea. Why was only the former explosive?

The answer lies in the amount of water and other gases dissolved in magma, about 1 percent of the weight of the molten rock in the Surtsey lavas. When this type of magma from depth is erupted into water, it quickly forms a chilled, glassy rind. At the low pressure in shallow water, most gases boil out of the molten rock and expand greatly in volume. This expansion fractures the glassy rind of the venting magma and gives the seawater access to a large surface area of finely broken but still hot rock. The steam generated by this contact water, added to the steam and other gases boiling out of the magma, creates the explosions.

Lava that has erupted from a crater on land and has then flowed through tubes to the sea has already lost most of its dissolved gases. As it flows into the sea, it forms sacklike bodies of lava with chilled, glassy rinds; these bodies are called *pillow lava* (Figure 19). These "pillows" do not explode from the expansion of internal gases because those gases have already boiled away on land. A relatively small surface area loses heat to the surrounding water, and some steam is formed without explosions.

This process suggests that lavas emitted from vents in deep water—deep enough that the dissolved gases cannot boil and expand—also quietly form pillow lavas or sheetlike flows. Because the contact water is under too much pressure to boil, no steam forms. The violence of fire under the sea is thus controlled by depth. The commonsense notion that fire and water do not mix originates in the limited view of our low-pressure surface environment. Of course, the word *fire* is used loosely. In this context, "fire" means "hot and glowing," not the combustion of a fuel with oxygen to produce flames. Deep under the ocean, fire and water find a way to exist side by side.

The critical depth below which submarine lava is not explosive depends on the dissolved gas content and the temperature of the magma, but for basaltic lava with a low gas content, that depth is only about 30

A

B

19 Pillow lava formed by volcanic flows under water makes up some of the most widespread rocks on Earth. A. Pillow lava formed beneath the prehistoric glacier ice on Iceland. B. Young pillow lava on the Mid-Atlantic Ridge 2700 meters below the ocean surface. (B, photograph taken from the submersible *ALVIN* by W. B. Bryan, Woods Hole Oceanographic Institution.)

20 The submersible *ALVIN* in the rift valley of the Mid-Atlantic Ridge. This artist's conception shows *ALVIN* exploring an open fracture cutting across young pillow lava at a depth of 2600 meters. (Painting by the National Geographic Society.)

meters below sea level. Most of the oceanic ridges where the seafloor is spreading and new oceanic crust is forming are more than 2000 meters below sea level. Pillow lavas and sheet flows thus form much of the pavement of the spreading seafloor. Rocks dredged from the ridges by oceanographic ships and sampled on recent dives by scientists in special submersibles confirm this theory (Figure 20).

Jim Moore, one of the divers who descended 3000 meters into the valley along the crest of the Mid-Atlantic Ridge, describes the scene as a strange, silent land of pillow lava hills and yawning cracks, with a dusting of sedimentary mud. The pillow lava forms steep-sided hills 20 to 30

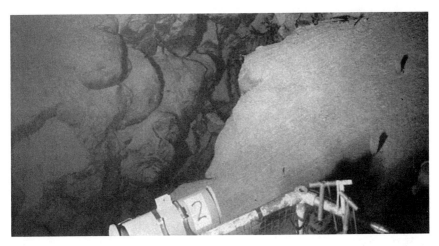

21 View down into an open fracture in the rift valley of the Mid-Atlantic Ridge. The fracture cuts across sediment-covered pillow lava and was evidently formed by the forces separating the plates. This photo was taken at a depth of about 2700 meters during Project FAMOUS (French–American Mid-Ocean Undersea Study) in 1974. Similar above-sea-level fractures cut through lava flows in the rift zones of Iceland. (Photograph by W. B. Bryan, Woods Hole Oceanographic Institution.)

meters high and 0.5 to 1 kilometer wide near the center of a 4-kilometer-wide rift valley. These hills appear fresh and recently erupted, clearly marking the very axis of the spreading oceanic ridge system.

On the valley floor on either side of the hills, many of the lavas look older. They are being covered with loose sedimentary debris that is slowly raining down from the ocean surface above. Most of the sediment is composed of microscopic shells of calcium carbonate from floating plants and animals. The relative ages of submarine lava flows can be established by the thickness of the sedimentary mud that slowly accumulates with time (Figure 21).

Deep cracks parallel to the rift valley cut across the older lavas. These cracks result from the pulling apart of the entire oceanic ridge system (Figures 22 and 23). The steep walls that form the sides of the rift valley appear to be normal faults along which the valley has dropped down about 500 meters relative to the ridge. For more than 1000 kilometers on either side of the rift valley, the Mid-Atlantic Ridge slopes generally downward over rugged and fractured volcanic terrain into deep basins 4000 to 5000 meters below sea level.

The East Pacific Rise spreads at a faster rate than the Mid-Atlantic Ridge; its magnetic stripes indicate separation rates of 6 to 17 centime-

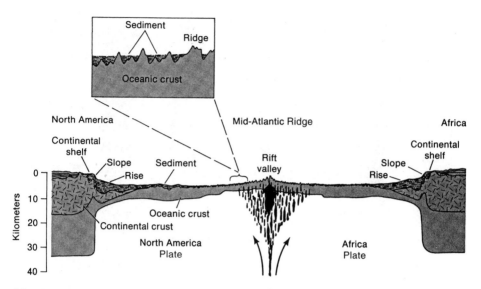

22 A cross section of the Earth beneath the Atlantic Ocean basin. The continents of North America and Africa are moving away from the Mid-Atlantic Ridge with the spreading seafloor. Sediments accumulating in the valleys between the rugged submarine mountains slowly bury those peaks. (After B. C. Heezen, in F. Press and R. Siever, *Earth*, 2nd ed., p. 262. Copyright © 1978, W. H. Freeman and Company.)

ters per year. Recent exploration of this rift system has revealed some interesting similarities and differences between Atlantic and Pacific seafloor spreading (Figure 24).

Exploratory dives in the submersible named *ALVIN*, operated by the Woods Hole Oceanographic Institution, were concentrated in an area just south of Baja California, at a depth of 2500 meters. There the rift axis is only about 1 kilometer wide, and the East Pacific Rise spreads at a rate of 6 centimeters per year. Besides young pillow lava, the divers found some smooth-surfaced basaltic lava flows and solidified lava lakes. A fracture zone with fissures generally parallel to the rift system extends 0.5 to 2 kilometers on each side of the youngest lavas. Major fault scarps, some as high as 70 meters and facing in toward the rift axis, extend outward from the spreading center. The locations of earthquakes and seabottom evidence of active faulting indicate that the spreading is concentrated within a 20-kilometer-wide zone, mostly near the central axis. Beyond this zone, the plates continue to move passively away on their long journey beneath the Pacific Ocean.

Submarine hot springs were also found on this expedition. Exotic life-forms and metal sulfide deposits associated with these hydrothermal vents are among the most exciting scientific discoveries of the past two

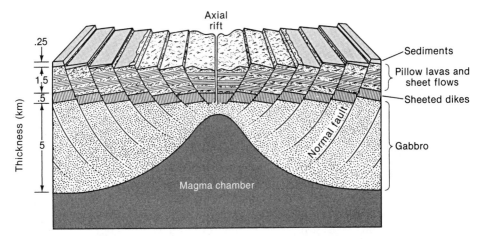

23 A schematic cross section of a spreading mid-ocean ridge. Active submarine volcanism is concentrated in a narrow area along the rift axis. Faulting forms the down-dropped blocks of older volcanic rock on both sides of the spreading axis. Magma forms as the plates diverge and mantle rock rises and melts with the decrease of pressure. It collects in a chamber below the spreading center. Along the borders of the magma chamber, a coarse crystalline rock of basaltic composition called *gabbro* is formed. At the top of the chamber, the magma rises as the plates diverge and cools as vertical dikes. At the surface, lava flows out and hardens in the form of sheets and pillows. As the new crust moves away from the spreading center, a layer of sediments is deposited on it. The mature crust has a layered structure from the top down: sediments, sheet and pillow lavas, dikes, and gabbros. (From Jean Francheteau, "The Oceanic Crust. East Pacific Rise." Copyright © 1983 by Scientific American, Inc. All rights reserved.)

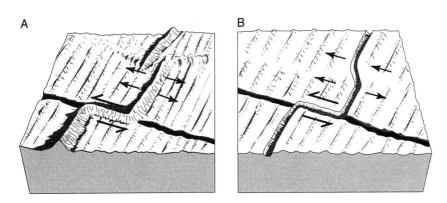

24 Comparison of the topography of the axial areas of a fast-spreading mid-ocean ridge (A, the East Pacific Rise) and a slow-spreading ridge (B, the Mid-Atlantic Ridge). The axis of a fast-spreading ridge is an elevated zone, possibly supported by the buoyant force of a shallow magma chamber beneath the axis. This axis is often offset 50 kilometers or more by major transform faults (A, bottom). Smaller offsets (A, top) are formed by overlapping ridge segments. The axis of a slow-spreading ridge is a valley. It is also offset by transform faults (B, bottom) or bends along the axis (B, top). (From Ken Macdonald and Paul Fox, "The Mid-Ocean Ridge." Copyright © 1990 by Scientific American, Inc. All rights reserved.)

decades. Warm springs (20°C) were first observed during dives on the Galápagos Ridge in 1977, and in 1979 extremely hot springs (350°C) were observed during dives at 21° north latitude on the East Pacific Rise. Bruce Luyendyk of the University of California at Santa Barbara was one of the scientists on the discovery dives, and his own words best describe the experience:

> The scene was like one out of an old horror movie. Shimmering water rose between the basaltic pillows along the axis of the neovolcanic zone. Large white clams as much as 30 centimeters long nestled between the black pillows, while crabs scampered blindly across the volcanic terrain. Most dramatic of all were the clusters of giant tube worms, some of them as long as 3 meters.

On a subsequent dive, the scientific submariners found another impressive sight: "extremely hot fluids, blackened by sulfide precipitates, were blasting upward through chimneylike vents as much as 10 meters tall and 40 centimeters wide. We named the vents 'black smokers'" (Figure 25 and Plate 19).

In the years since, several other undersea expeditions have discovered these strange oases of life and sulfide metal deposits at scattered places along the axes of spreading rifts in both the Atlantic and Pacific oceans. They are apparently common, but not continuous, features of the entire world rift system, and are most likely to be found where recent and frequent volcanic activity provides the heat and gases needed to sustain their exotic food chain and mineral deposition.

The significance of these strange colonies of life on the dark seafloor and the associated mineral deposits is just beginning to be appreciated fully by biologists and geologists. The bacteria that feed the food chain of strange worms, clams, and crabs live on volcanic heat and gases; they do not need sunlight (Figure 26). Before this discovery, the sun and photosynthesis were considered to be the essential suppliers of energy to life on Earth.

The very latest twist in this fantastic discovery is proof that some of the bacteria that sustain this food chain are not bacteria at all, but a previously unknown form of life. These puzzling one-celled organisms, named Archaea, were found to have genes distinct from those of any other known forms of life. Craig Venter—of the Institute of Genomic Research and leader of the team of scientists that mapped the 1738 genes of Archaea that were cultured from submarine vent samples—remarked that "two thirds of its genes don't look like anything we've ever seen in biology before." Some of the genes of the more familiar third are simi-

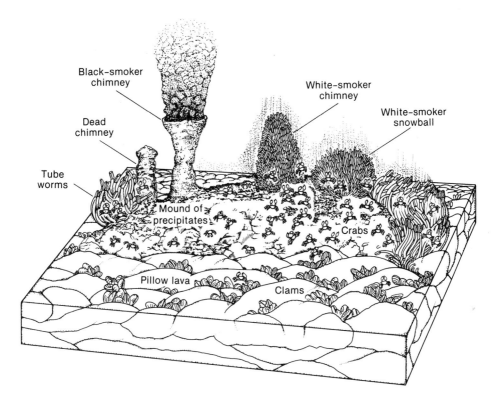

25 Idealized scene of living creatures and metal sulfide precipitates at a submarine hot spring 2500 meters deep along the rift zone of the East Pacific Rise. The white-smoker chimney is made of burrows of Pompeii worms. White, cloudy fluids emitted by the white smokers have temperatures up to 300°C. The hottest water, up to 350°C, pours from the black smoker and precipitates a dark cloud of metal sulfide particles as it is cooled by the surrounding seawater. (From K. C. Macdonald and B. P. Luyendyk, "The Crest of the East Pacific Rise." Copyright © 1981 by Scientific American, Inc. All rights reserved.)

lar to those in humans, and others are similar to those in bacteria. The Archaea of the submarine oases thrive in total darkness on a diet of carbon dioxide, hydrogen, nitrogen, and volcanic heat. The main by-product of their metabolism is methane.

The metallic sulfide deposits are an equally important discovery. They precipitate where seawater chills the submarine hot springs, forming mounds and crusts that are rich in iron, copper, and zinc. As the seafloor separates, these deposits are spread into extensive layers of potentially great economic value (more on this in Chapter 16). One of the important lessons from the submarine exploration of the mid-ocean ridges

26 Tube worms and other unusual life-forms flourish near submarine hot springs. This photograph shows clusters of large tube worms, some as long as 2 meters, waving in warm water at the margins of hydrothermal vents. Chemosynthetic bacteria form the base of this sunless food chain, which also supports clams and crabs. (Photograph taken at 21° north latitude on the East Pacific Rise during the RISE Expedition, F. N. Speiss, Scripps Institution of Oceanography, 1979, and provided by William Normark, U.S. Geological Survey.)

is the value of serendipity—while searching for one thing, other things are discovered, which are sometimes even more important than what was originally sought.

These recent submarine explorations and new detailed maps of the seafloor provide a good picture of the topography and structure of the mid-ocean ridges. In the faster-spreading ridges, such as the East Pacific Rise, the axis of the system is a ridge several hundred meters high and 5 to 20 kilometers wide. This spreading axis is interrupted every few hundred kilometers by transform faults that offset the axis by many tens of kilometers (see Figure 24A). Within the segments separated by transform faults are smaller undulations and offsets of the central ridge crest, about every 100 kilometers. These features may be related to segmentation of the magma supply reservoirs (Figure 27).

In the slower-spreading ridges, such as the Mid-Atlantic Ridge, which diverge at about 3 centimeters per year, the axis of the system is a rift valley a few kilometers deep and 30 to 40 kilometers wide. The depth to the magma reservoirs is apparently greater than in the fast-spreading ridges, and these chambers seem to be spaced about every hundred kilometers. Recent explorations by the Woods Hole Oceanographic Institu-

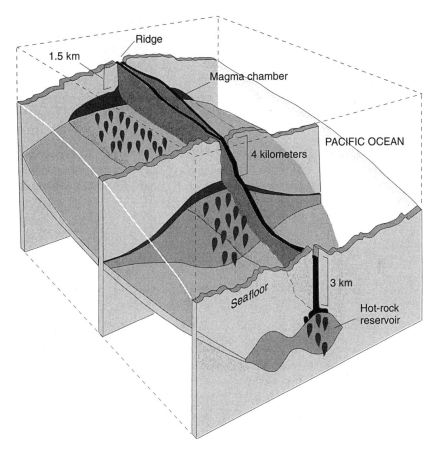

27 A schematic view of the rocks, structures, and magma chamber thought to occur beneath the East Pacific Rise. (From Perfit and Chadwick, American Geophysical Union, 1998.)

tion of the Gakkel Ridge, which extends for 1800 kilometers beneath the Arctic Ocean, have established a third category of divergent margin that spreads even more slowly—less than 20 millimeters per year. Here the plates are separating so slowly that the volcanism does not entirely cover the tear. Nevertheless, these expeditions mapped 12 previously unknown underwater volcanoes and a large area of undersea hydrothermal vents. Overall, the great mid-ocean ridge system is like the seam of a baseball. It is a nearly continuous feature, broken only by names as it extends through the ocean basins of the world.

One of the main debates concerning rift volcanoes centers on whether they are a cause or an effect of the rifting apart of the oceanic ridges. Does the intrusion of volcanic rock push apart the plates, or does

28 Aerial photograph of fractures in the rift zone of Iceland. These fractures mark the continuation of the crest of the Mid-Atlantic Ridge across Iceland. The lava flows in this area are less than 6000 years old. Summer homes dot the edge of the lake formed by the central valley of the rift.

the pulling or dragging apart of the plates form cracks through which the magma can rise to the surface? Most experts now favor the latter view, although the answer may be both.

During the past 35 years, Icelandic scientists have been measuring the spreading in their country, using sensitive surveying instruments that can detect changes of a few millimeters over distances of many kilometers. The results of these studies indicate that Iceland slowly widens about 2 centimeters per year, across distances of hundreds of kilometers, without earthquakes or volcanic eruptions. When the tension reaches the breaking point, the rift fractures (Figure 28). A rifting event like this takes place in a few hours or days. Open cracks form parallel to the rift, earthquake swarms jolt the local area, and lava erupts from some of the fissures.

We were lucky enough to be in Iceland when one of these rifting episodes occurred on September 7, 1977. An earthquake swarm at Krafla Volcano was followed in rapid succession by a brief but spectacular volcanic fissure eruption, with the ground cracking parallel to the rift (Figure 29). In a zone about 2 kilometers across and 20 kilometers long, the

29 A gaping fracture formed during an earthquake swarm in northern Iceland on September 7, 1977. This rifting event was related to an eruption of Krafla Volcano. The cone in the background was formed during a prehistoric eruption of Krafla.

30 Incandescent lava was briefly ejected from this geothermal steam well in Iceland during the eruption of Krafla Volcano on September 7, 1977. A fracture somewhere between the surface and the 1200-meter depth of the well broke the well casing and allowed a small amount of molten rock to enter the producing well. The steam lifted the magma spray to the surface, where it melted and abraded its way through the pipe. The few tons of small lava bombs that were ejected form the granular surface material. This is the only known case of a volcanic vent produced by drilling.

cracking formed open fractures as wide as 20 centimeters with vertical offsets as high as 1 meter. All this occurred between 3 P.M. and midnight.

One of the cracks cut through a producing geothermal steam field. Some magma was injected deep along this fracture, and one of the steam wells erupted small bursts of incandescent lava. The erupting lava cut through the heavy metal pipe at the wellhead and showered a few tons of frothy lava fragments like black popcorn around the well. This is the only known case in which lava has erupted from a man-made vent (Figure 30).

A survey line about 4 kilometers long that crossed the fracture zone lengthened by more than 1 meter, but survey lines on the flanks of the new fracture zone showed contraction. The picture of rifting that emerges is one of slow stretching that takes place over a wide zone for hundreds of years, followed by rupture that concentrates the deformation into a narrow central zone and allows the flanks to contract back to an unstretched condition. It is like pulling a rubber band slowly and then cutting it in the middle; all the stretching that had slowly accumulated is suddenly concentrated in the cut, and the sides contract. Perhaps a sudden tear in a piece of elastic cloth when it is slowly stretched would be a better analogy, since its dimensions fit the plates more closely than does a rubber band.

The length of the tear in a rifting episode is tens of kilometers long, only a very small fraction of the length of the entire rift system. Rifting episodes are infrequent in any one place, perhaps occurring only once every few hundred years. In any given year, however, rifting probably takes place somewhere along the length of the oceanic ridge system.

The presence of magma under pressure beneath the rift zones probably aids the rupturing process, just as a wedge speeds the felling of a leaning tree. The processes of rifting and volcanism are so interrelated that it is difficult to sort out cause and effect—reminding us of Samuel Butler's remark that "a hen is just an egg's way of making another egg."

There is no doubt, however, that great chains of volcanoes, mostly submarine, circle the Earth along the crests of the oceanic ridges. They owe their origin to the still mysterious processes that spread the plates. Surtsey is only a visible tip of this vast hidden activity.

Links

National Oceanic and Atmospheric Administration (NOAA)
   http://www.noaa.gov/ocean.html

Columbia University Ocean-Ridge Project
   http://ocean-ridge.ldeo.columbia.edu/

Ocean Ridge Maps
http://ocean-ridge.ldeo.columbia.edu/general/html/home.html

Woods Hole Oceanographic Institute, Education
http://www.divediscover.whoi.edu/about.html

Seafloor Hydrothermal Vents, Woods Hole, 2004
http://www.whoi.edu/home/index_education.html (type "Hydrothermal Systems" in search box)

Juan de Fuca Ridge (NOAA), 2003
http://www.pmel.noaa.gov/vents/nemo/education.html

Links to other seafloor sites, 2003
http://www.pmel.noaa.gov/vents/nemo/links.html#vents

Submarine Volcanic Eruptions
http://www.pmel.noaa.gov/vents/geology/submarine_eruptions.html

Sources

Cone, Joseph. *Fire under the Sea*. New York: Morrow, 1991.

Macdonald, Ken C. "Linkages Between Faulting, Volcanism, Hydrothermal Activity and Segmentation on Fast-Spreading Centers." In *Faulting and Magmatism at Mid-Ocean Ridges*, Geophysical Monograph 106. Washington, DC: American Geophysical Union, 1998.

Macdonald, Ken C., and Bruce P. Luyendyk. "The Crest of the East Pacific Rise." *Scientific American*, May 1981.

Nicolas, Adolphe. *The Mid-Ocean Ridges*. New York: Springer-Verlag, 1994.

Perfit, M. R., and W. W. Chadwick. "Magmatism at Mid-Ocean Ridges." In *Faulting and Magmatism at Mid-Ocean Ridges*, Geophysical Monograph 106. Washington, DC: American Geophysical Union, 1998.

Solomon, S. C., and D. Toomey. "The Structure of the Mid-Ocean Ridges." *Annual Review of Earth and Planetary Sciences* 20 (1992): 329–364.

CHAPTER 4

Mount St. Helens

31   On May 18, 1980, a giant explosive eruption and avalanche destroyed in moments the entire region surrounding Spirit Lake below Mount St. Helens. Not a tree was left standing; devastation was everywhere. But Earth abides; life will return to Spirit Lake over the centuries, as it did before. (Photograph by the U.S. Geological Survey.)

Men argue; Nature acts.
—Voltaire (1694–1778)

The violent eruption of Washington State's Mount St. Helens in May 1980 was one of the world's most closely monitored—and most fully documented—volcanic events (Figures 32 and 33). Volcanologists had been watching the mountain with keen interest, alerted by two months of mild but ominous activity that signaled the awakening of the sleeping giant (Figure 34).

At 8:32 A.M. on May 18, Mount St. Helens was shaken by a magnitude 5 earthquake, centered 1 to 2 kilometers beneath its north flank. Geologists Keith and Dorothy Stoffel, who were flying in a light plane about 400 meters above the summit at just that moment, saw several small icefalls start down the steep crater walls. About 15 seconds later they were the closest witnesses to the onset of the largest landslide in recorded history, closely followed by a huge volcanic eruption (Figure 33).

"The whole north side of the summit crater began to move instantaneously as one gigantic mass. The nature of movement was eerie; not like anything we'd ever seen before," Dorothy recalled. "The entire mass began to ripple and churn without moving laterally. Then the whole north side of the summit started moving to the north along a deep-seated slide plane" (Figure 35B).

Seconds later, a massive explosion shook the mountain. From the Stoffels' viewpoint, the initial explosion cloud seemed to mushroom sideways to the north and plunge down the slope, but in the plane the Stoffels neither felt nor heard the explosion. Realizing the enormous size of the eruption, survival became their first concern. They dove at full throttle to gain speed, but the expanding cloud appeared to be gaining on them; by turning south, they finally outran it. Behind them, rapidly growing ash clouds thrust north and northwest. To the east, the black clouds rose into billowing mushroom shapes, and lightning bolts thousands of meters high shot through them. Half an hour later, the Stoffels, shaken but safe, landed at the Portland, Oregon, airport.

In the avalanche they had witnessed from the air, nearly 3 cubic kilometers of crushed rock and glacier ice plunged down into Spirit Lake and the North Fork of the Toutle River. Exploding steam helped fluidize

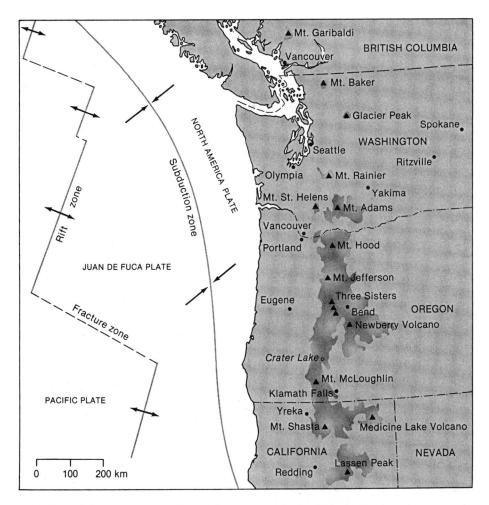

32 Map of the northwestern United States, showing the Cascade Range volcanoes and their relationship to the subduction of the Juan de Fuca Plate beneath the North America Plate. Submarine volcanoes and hot springs have recently been discovered along the rift zone that forms the western edge of the Juan de Fuca Plate. The dark gray areas are volcanic rocks less than 2 million years old. (Data from the U.S. Geological Survey.)

the avalanche, and it accelerated rapidly to velocities of 250 kilometers per hour. One lobe of the gigantic mass plowed through the west arm of Spirit Lake and into the valley to the north of the lake. The momentum of another lobe carried it over a ridge 360 meters above the floor of the valley that it had just swept across, but the bulk of the fluidized debris was funneled down the valley of the Toutle River for 21 kilometers, form-

33 This remarkable sequence of photographs of the explosion of Mount St. Helens was taken by Keith Ronnholm, a geophysicist from Seattle, Washington, on the morning of May 18, 1980. He was camped near Bear Meadows, 18 kilometers northeast of the peak. After taking the pictures, Ronnholm was enveloped in a blackout of heavy ash and pumice fall, but he was just beyond the area of total devastation. The top photograph, taken 40 seconds after the great avalanche began, shows the huge landslide blocks being overtopped by the subsequent explosion. The bottom photograph, taken 20 seconds later, shows the rapidly growing explosion cloud thrusting northward from Mount St. Helens. (Photographs by Keith Ronnholm, © 1980.)

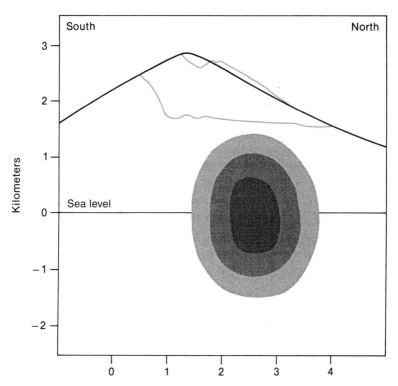

34 This cross section of Mount St. Helens shows the changes associated with the great 1980 eruption. The solid line at the top indicates the surface as of 1979. The upper light gray line shows the crater and bulge that developed between late March and early May 1980. The shaded area indicates the region where the thousands of earthquakes that accompanied the growing bulge occurred; the darker the shading, the higher the density of earthquake locations. The lower light gray line shows the crater after the May 18 eruption. (Data from the U.S. Geological Survey and University of Washington Department of Geophysics.)

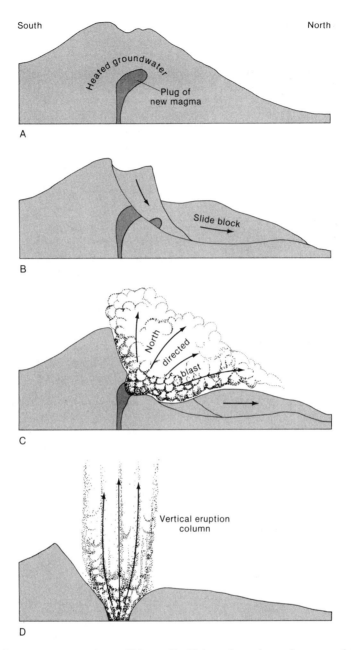

35 These schematic cross sections of Mount St. Helens show the early stages of the May 18, 1980, eruption. A. The shallow plug of magma injected beneath the volcano has caused the high north face to bulge outward and has heated the groundwater. Minor explosions of heated groundwater have excavated a small summit crater. B. The magnitude 5 earthquake caused by the continuing injection of the magma plug shakes loose the oversteepened north flank of the mountain into two or more massive slide blocks. C. As the superheated groundwater and magma plug are suddenly depressurized, they explode in a huge steam blast that sweeps northward for many kilometers. D. After the avalanche and blast have exposed the main magma conduit, gases dissolved in the magma continue to boil out in continuous vertical jetting explosions. (Modified from J. G. Moore and W. C. Albee, U.S. Geological Survey Professional Paper 1250, 1981, p. 132.)

36 View from the west of Mount St. Helens, showing the huge avalanche deposit in the valley of the North Fork of the Toutle River. (U.S. Geological Survey photograph by R. M. Krimmel, June 30, 1980.)

ing a hummocky deposit 1 to 2 kilometers wide and up to 200 meters thick (Figure 36).

Superheated groundwater flashed into steam as the great avalanche of rock and ice suddenly released the pressure in the volcanic edifice. At the same time, dissolved gases began exploding from the shallow magma body that had recently pushed up into Mount St. Helens (Figure 35C). The steam blast, magma explosion, and giant avalanche combined to form a lateral blast of dense, debris-filled steam clouds as hot as 300°C, which surged northward from the breached volcano at speeds as high as 1000 kilometers per hour. This blast of steam and its fluidized charge of

37 Trees blown down by the hot, debris-filled, hurricane-force winds in the Mount St. Helens steam-blast cloud fell into streamlined rows over hundreds of square kilometers. These trees are large Douglas firs and other conifers, 1 to 2 meters in diameter and 40 to 60 meters long. (Photograph by Katia Krafft.)

volcanic rock fragments devastated nearly 600 square kilometers of mountainous terrain northwest, north, and northeast of Mount St. Helens. The ground-hugging black clouds rolled over four major ridges and valleys, extending as far as 30 kilometers from their source.

The destruction was complete. For the first few kilometers, entire trees as large as 2 meters in diameter were uprooted and swept away with the roiling explosion cloud. Beyond this was a blowdown zone 10 to 15 kilometers wide where prime fir trees were snapped off like matchsticks (Figure 37). At the outer limits of destruction, trees were still standing, but their needles had been scorched beyond recovery.

The blowdown zone looked as if a great shock wave or concussion front had knocked down the trees in a pattern radial to the exploding peak, but the evidence showed that this was not what had happened. Survivors near the edge of the devastated area heard only a moderately loud explosion or roaring sound two to three minutes before the black cloud with its hot hurricane-force winds descended upon them. The average velocity of the growing front of the steam-blast clouds was below the speed of sound. On closer inspection, the pattern of downed trees showed tur-

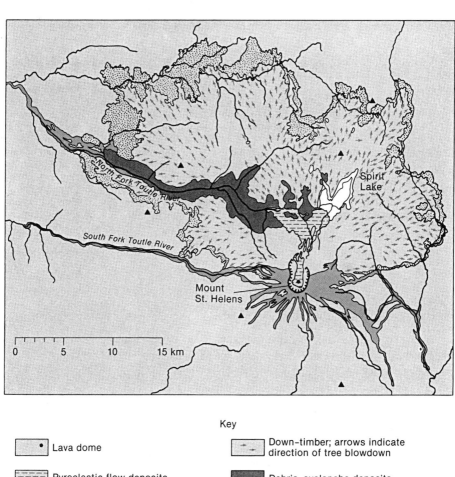

Key

Lava dome	Down–timber; arrows indicate direction of tree blowdown
Pyroclastic flow deposits	Debris-avalanche deposits
Mudflow deposits and scoured areas	▲ Mountain summits
Scorch zone	

38 Map of the region surrounding Mount St. Helens, showing the effects of the May 18, 1980, eruption and the subsequent lava dome. The devastated area is more than 500 square kilometers. The size of Spirit Lake was increased by the eruption, as seen in both the old and the new shorelines. (Adapted from the U.S. Geological Survey.)

bulent eddies and curving streamlines. Near the edge of the devastated area, most trees were blown down in down-valley directions, even turning toward the source of the surging clouds (Figure 38).

Gravity apparently provided energy to the dense, fluidized mass as the initial steam-blast energy waned. As their turbulent internal winds abated, ash and rock debris from the dense explosion clouds settled to the surface in deposits that ranged in thickness from about 1 meter to a few millimeters. Angular fragments of old volcanic rocks and hot but recently solidified magma were carried in the steam-blast clouds as far as 15 kilometers from the exploding peak. Trees were charred and blackened on their sides facing the blast for 7 kilometers on the northwest and north, and as far as 18 kilometers on the northeast flank of Mount St. Helens.

The worst was over in the first five minutes, but a vertical eruption column roared on (Figure 35D), reaching a maximum height of about 26 kilometers at 9 A.M. and then diminishing in altitude to about 14 kilometers for much of the day. The source of this almost continuously exploding and uprushing column of gas and ash was the effervescing shallow magma body, which was being progressively cored out to greater and greater depths. The abrasive uprush continued to enlarge the horseshoe-shaped crater that initially had been formed by the avalanche and lateral blast.

High-altitude winds blew all day to the northeast, and ash began falling on central Washington towns by midmorning. At Yakima (150 kilometers away), the first ashfall was a sand-sized "salt and pepper" layer composed of dark rock fragments and lighter-colored feldspar crystal fragments, overlain by a thicker, silt-sized layer of volcanic glass particles. Thirty kilometers north of Yakima, the maximum ashfall was about 20 millimeters. Eastward the deposits of fine ash thickened, reaching more than 70 millimeters near Ritzville, Washington, 330 kilometers away from Mount St. Helens. Here the texture of the ash was like that of talcum powder. Although the ash deposits thinned to 5 millimeters near Spokane, 430 kilometers northeast of the volcano, by 3 P.M. visibility there was reduced to 3 meters in near darkness. A trace of ash fell on Denver at about noon on May 19; the ash cloud crossed the United States in three days.

Mudflows were another major result of the eruption. These flows consisted of a slurry of volcanic ash and fine rock particles mixed with water, which had the consistency of wet cement. The ash blanket near the mountain and the crushed rock in the avalanche deposit provided the solid matter; the extra water probably came from several sources— melting snow and ice, the water displaced from Spirit Lake and the North Fork of the Toutle River, water from the compaction and settling of the saturated avalanche deposits, and condensing steam.

The highest flood crest in the North Fork of the Toutle River arrived at the gauging station that evening and destroyed it. High mudflow marks on trees indicated a flood stage of 16 meters, 9 meters higher than any flood previously recorded on the Toutle River. Downstream, mudflow deposits clogged the channels of the Cowlitz River and caused severe shoaling of the navigation channel in the Columbia River.

Sometime after the initial avalanche and steam-blast eruption, *pyroclastic flows*—fluidized emulsions of hot (though not molten) rock fragments and volcanic gases—began rushing down the north slope of Mount St. Helens through the breach in the newly formed crater (Figure 39). Fine ash and pumice blocks formed dense flows that poured from the crater beneath the rising ash cloud. Successive flows churned down the north slope at speeds up to 100 kilometers per hour, covering the earlier landslide and blast deposits and reaching the south edge of Spirit Lake. Contact of these hot pyroclastic flows with water caused secondary explosions, sending steam and ash clouds 2 kilometers high.

The total scope of disaster was staggering: 57 people killed and more than $1 billion in losses and damage—mostly to the timber industry. Perhaps the greatest damage was psychological: for the people of the northwestern United States, the image of the Cascade volcanoes had changed from silent guardian peaks to potential killers (Figure 40).

Although the suddenness, complexity, and scale of the May 18 eruption were not anticipated, the fact that a major eruption took place was not a complete surprise. In 1978 Dwight Crandell and Donal Mullineaux of the U.S. Geological Survey (USGS) issued a report about Mount St. Helens that concluded with the warning, "The volcano's behavior pattern suggests that the current quiet interval will not last as long as a thousand years; instead, an eruption is more likely to occur within the next hundred years, and perhaps even before the end of this century." This forecast was based on their studies of the character, distribution, and ages of prehistoric eruption deposits from the volcano, as well as scattered accounts by explorers describing eruptions of Mount St. Helens between 1831 and 1857. Just two years after this forecast was published, the quiet interval ended.

Beginning on March 20, 1980, a swarm of small to moderate earthquakes centered beneath Mount St. Helens announced the volcano's awakening. A seismometer had already been placed on the west flank by University of Washington seismologists, and soon after the swarm started, additional seismometers were installed to record and locate the earthquakes more precisely. On March 25, within a 12-hour period, there were 47 earthquakes of magnitude 3 or greater at shallow depths under the volcano's north flank.

The first eruptive activity—small explosions of steam—began two days later, forming a new crater about 70 meters wide on the snow- and ice-covered summit. Large east–west cracks also developed across the summit area. A second small crater formed on March 29, and blue

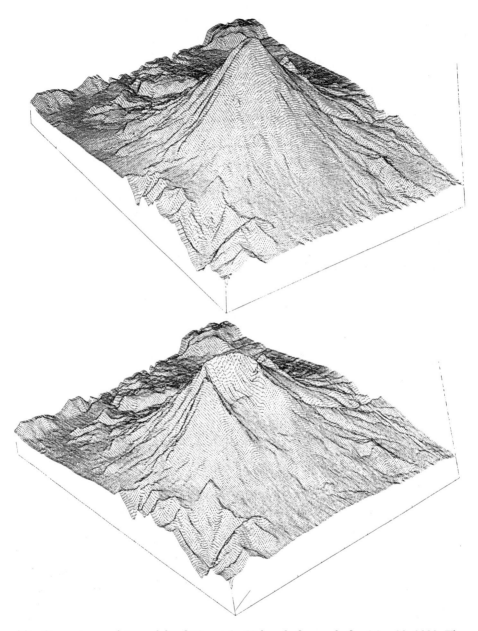

39 Computer-graphic models of Mount St. Helens before and after May 18, 1980. The avalanche and eruption removed the top 400 meters of the peak, forming a breached crater 2 kilometers wide, 4 kilometers long, and 700 kilometers deep. These views are from the northeast and exaggerate the relief. (Created by the U.S. Geological Survey and Dynamic Graphics, Inc.)

40 Before-and-after views of Mount St. Helens. The photograph at the top was taken in June 1970, the one at the bottom in May 1980. The pre-1980 summit elevation was 2950 meters, with the cone rising from a base with an elevation of about 1000 meters. The rim of the post-eruption crater has elevations ranging from 2400 to 2550 meters, and the crater floor elevation varies from 1800 to 1900 meters. (1980 photograph by Ross Hamilton, U.S. Geological Survey.)

flames—possibly burning hydrogen sulfide gas—were visible from the air at night. The next day, scientists counted 93 small eruptions of steam and ash.

On April 1 the seismographs recorded the first volcanic tremor, a more or less continuous ground vibration that is observed at many active volcanoes. The precise cause of volcanic tremor is not clear, but it presumably reflects the movement of magma or the rumbling formation and collapse of subsurface gas bubbles associated with the incipient boiling of groundwater or gases dissolved in magma.

The small steam and ash explosions continued, some as single explosions and others as pulsating jets lasting for hours. While the craters grew into a single oval pit 500 meters by 300 meters across and 200 meters deep, columns of gas and ash climbed to 3 kilometers above the summit. The ash consisted of fragments of old volcanic rock, and the gas emissions were largely steam mixed with carbon dioxide, sulfur dioxide, hydrogen sulfide, and hydrogen chloride. The small explosive eruptions during this period apparently took place when groundwater high in the volcanic cone was heated above its subsurface boiling point and suddenly flashed into steam, much as a geyser does, but with enough energy to tear loose rock fragments and blast out a crater.

As the high level of seismic activity continued—about 50 earthquakes of magnitude 3 or greater per day—another ominous sign became apparent. Geologists who were on the mountain taking measurements noticed that the profile of its north flank seemed to show distortion (see Figure 34). When they compared their view with a photograph in a climbing guide taken from the same location, it was evident that a major deformation was taking place. Careful measurements showed that a bulge was growing upward and outward, creating a spreading network of large cracks in the cover of snow and ice. Maps made by the USGS from aerial photographs taken on April 12 disclosed that the bulge was 2 kilometers in diameter and had already moved up and out by as much as 100 meters.

Repeated ground surveys made in late April and early May showed that the bulge was continuing to expand northward at a rate of about 1.5 meters per day. This rapidly deforming area was directly over the center of the earthquake zone 2 kilometers below. Scientists on the site agreed that the continuing earthquake swarm and surface bulging were signs that magma was being injected at a shallow depth beneath Mount St. Helens. They concluded that if the intrusion of magma continued at this high rate, a significant eruption was likely; the main questions were how soon it might occur and how violent it might be.

Small steam and ash explosions continued through the first half of May, and the rate of seismic and deformation activity continued unabated. On May 15, 16, and 17, there were no eruptions, and although the earthquakes and bulging continued, their rates showed no change. May 18 dawned as a beautiful spring day, but at 8:32 that morning, the scientists' worst fears became a reality.

Among the dead and missing that day was David Johnston, a USGS scientist who was measuring the growing bulge from a ridge 10 kilometers north of the volcano's summit. The scale of the eruption—especially the great area devastated by the lateral blast, which was unprecedented in the prehistoric eruption record of Mount St. Helens—was unexpectedly large. The suddenness with which the avalanche broke open the hot groundwater and magma system beneath the summit of the volcano was responsible for much of its violence. Had the unroofing of the shallow magma injection occurred in a slower, more piecemeal way—for example, in a series of increasingly larger explosions over a period of several days—the area devastated by the eruption would have been much smaller.

On five more occasions through the summer and early fall of 1980, smaller explosive eruptions of Mount St. Helens generated ash clouds and pyroclastic flows (Plate 14). Twice, *lava domes*—thick extrusions of viscous lava—were pushed up over the vent area in the crater, but then were blown out by explosions. Between late 1980 and 1986, a large lava dome episodically squeezed up over the vent area, forming a dome complex about 1 kilometer in diameter and 300 meters high. These episodes of rapid dome growth typically lasted for several days, separated by quiet intervals of a few months to several years.

Each of these rapid dome growth episodes was preceded by several days of increasing seismic activity and increasing rates of ground deformation at the margins of the dome. Continuous seismic and deformation monitoring allowed scientists to predict these episodes with accuracy. Because the amount of growth in each episode was roughly proportional to the length of the quiet interval that preceded it, the amount of lava injected in each new episode also was predictable.

The last dome-growth episode of the series occurred in October, 1986. In the years that followed, Mount St. Helens's activity began to recede into history. Then, 18 years later, to almost everyone's surprise, the mountain stirred again. On October 1, 2004, a small steam and ash eruption from the crater heralded the beginning of a new episode of dome growth that continues at the time of this writing (March 2005). Seismic unrest preceding the eruption began on September 23 with a small swarm of shallow earthquakes beneath the crater. These tremors increased in both rate of occurrence and magnitude over the next few days, and were accompanied by uplift of glacier ice on the crater floor on the south side of the 1986 lava dome.

41 Aerial view of Mount St. Helens looking west. The steaming 2004–2005 lava dome is on the south (left) of the snow-covered 1980–1986 dome complex. (Photograph by John Pallister, U.S. Geological Survey.)

Five small explosions of steam and ash occurred during October 1–6 as the uplift continued (Plate 15). The eruptions hurled old lava blocks as far as a kilometer, and ash rose about a kilometer above the crater. Mount St. Helens was definitely no longer dormant. The first new lava pushed up in spines from the deformed area on October 11, and in late October a large whaleback-shaped extrusion of lava gradually emerged. By December the area of uplifted glacier ice and talus and the new lava dome had grown to a combined volume of 30 million cubic meters, and by January the 2004–2005 dome was 1600 feet long, 1000 feet wide, and 550 feet high on the south side of the 1986 dome (Figure 41).

Measurements indicate that the new lava dome is 65 percent silica (crystal-rich dacite) and, in comparison to the 1980s dome, is relatively low in temperature (850°C), gas-poor, and of high viscosity. Gas emissions accompanying the new dome growth as of early January 2005 ranged from 800 to 2400 tons per day of carbon dioxide (CO_2) and 40 to 250 tons per day of sulfur dioxide (SO_2).

So what will the next move of this surprising volcano be? Geologists like to say that Mount St. Helens hasn't yet done all the dances she knows; even with all the new instrumentation that has become ever more sophisticated over the 25 years, no one can predict just what her next steps might be.

Links

Cascades Volcano Observatory, U.S. Geological Survey, 2004
http://vulcan.wr.usgs.gov/

Mount St. Helens, U.S. Geological Survey, 2004
http://vulcan.wr.usgs.gov/Volcanoes/MSH/framework.html

Mount St. Helens National Volcanic Monument, USDA Forest Service, 2004
http://www.fs.fed.us/gpnf/mshnvm/

Mount St. Helens VolcanoCam, USDA Forest Service, 2004
http://www.fs.fed.us/gpnf/mshnvm/volcanocam/

Mount St. Helens Slide Set, U.S. Geological Survey, 2004
http://vulcan.wr.usgs.gov/Volcanoes/MSH/SlideSet/ljt_slideset.htm

Pacific Northwest Seismograph Network, University of Washington, 2004
http://www.geophys.washington.edu/SEIS/PNSN/HELENS/

Mount St. Helens Maps and Graphics, U.S. Geological Survey, 2004
http://vulcan.wr.usgs.gov/Volcanoes/MSH/Graphics/framework.html

Sources

Carson, Rob. *Mount St. Helens: The Eruption and Recovery of a Volcano.* Seattle: Sasquatch Books, 1990.

Decker, Robert, and Barbara Decker. "The Eruptions of Mount St. Helens." *Scientific American*, March 1981.

Decker, Robert, and Barbara Decker. *Road Guide to Mount St. Helens.* Mariposa, CA: Double Decker Press, 2002.

Decker, Robert, and Barbara Decker. *Volcanoes in America's National Parks.* Hong Kong: Odyssey Guides, 2001.

Dzurisin, Dan et al. "Mount St. Helens Reawakens." *Eos* (*Transactions, American Geophysical Union*), January 18, 2005.

Lipman, Peter, and D. L. Mullineaux, eds. *The 1980 Eruptions of Mount St. Helens.* U.S. Geological Survey Professional Paper 1250, 1981.

Mullineaux, Donal. *Pre-1980 Tephra-Fall Deposits Erupted from Mount St. Helens, Washington.* U.S. Geological Survey Professional Paper 1563, 1996.

Pringle, Patrick. "Roadside Geology of Mount St. Helens National Volcanic Monument and Vicinity." Information Circular no. 88. Olympia: Washington Department of Natural Resources, 1993.

Thompson, Dick. *Volcano Cowboys: The Rocky Evolution of a Dangerous Science.* New York: Thomas Dunne Books, 2000.

Tilling, Robert, Lyn Topinka, and Donald Swanson. *Eruptions of Mount St. Helens.* Washington, DC: U.S. Government Printing Office, 1990.

CHAPTER 5

Mount Pinatubo, Philippines

42 View of Mount Pinatubo and the steam plume, looking west from Clark Air Base on June 6, 1991. The building in the foreground was the first Pinatubo Volcano Observatory (PVO). (From *Fire and Mud*, Newhall and Punongbayon, 1996, used with permission.)

In March 1991, people living on the upper slopes of Mount Pinatubo,
on the island of Luzon in the Philippines, began to feel an unusual number of small earthquakes. Most people at that time did not recognize that
the old, eroded 1745-meter-high mountain was a volcano. But on April
2, steam explosions marked the beginning of an eruption that was to climax on June 15 with the world's second largest volcanic eruption of the
twentieth century (Figure 42).

Compounding the danger from a newly restless volcano was the dense
population and the presence of two United States military bases in the
lowlands that surrounded the mountain (Figure 43). The Philippine Institute of Volcanology and Seismology (PHIVOLCS), headquartered
near Manila, less than 100 kilometers away, responded quickly to the
steam explosions and the increase in local earthquakes. PHIVOLCS also
invited several colleagues from the U.S. Geological Survey (USGS) to
join their investigations. Over the next two months, this joint investigative team issued eruption forecasts and evacuation plans that were to save
tens of thousands of lives. The priorities of the team were fourfold: establish a network of local seismometers to record the nature and locations of the increasing earthquakes; study the prehistoric volcanic deposits around Pinatubo to establish the age and severity of previous
eruptions; monitor the nature of the increasing volcanic activity and compare it with other historic eruptions—for example, that of Mount St. Helens; and somehow educate the people in the areas surrounding Pinatubo
about the nature and magnitude of the dangers they were facing.

This was a huge task for a small band of geologists and geophysicists.
Their success, and it was a great success, was due to the skill, energy,
persistence, and teamwork of the investigators. Their warnings, at first
unheeded, gained credibility when it became evident during the week
before the cataclysmic eruption that the awesome eruptive power of the
volcano was progressively increasing.

Pinatubo threatened a half-million people, and eventually hundreds
of thousands were evacuated. About 250 people were killed by the eruption, and a hundred more by the *lahars* (an Indonesian name for mudflows) that followed for years. The success of the eruption forecasts and

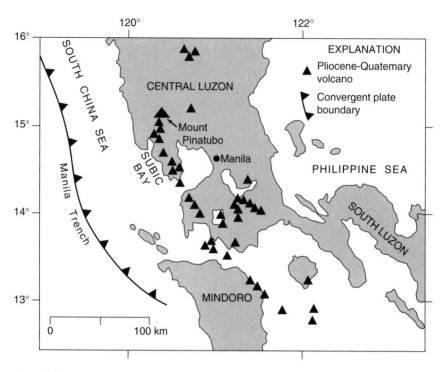

43 The location of Mount Pinatubo and other volcanoes in relation to the eastward-dipping Manila subduction zone in the Philippines. (See Figure 51 for the location of Clark Air Base.) (From *Fire and Mud*, Newhall and Punongbayon, 1996, used with permission.)

evacuations can be measured by the tens of thousands of people who heeded the warnings and survived.

This incredible story of nature's wrath, and of human endeavor to survive it, began with the initial steam blasts that occurred on the afternoon of April 2, 1991. On that day, a series of explosions blasted out a 1.5-kilometer-long, northeast-trending fissure in a sparsely populated jungle area on the north flank of Pinatubo (Figure 44). During and following the explosions, vigorous steam vents with strong sulfur odors burst forth on the upper slopes of the mountain. Two of these vents remained active during April and May, emitting plumes of steam that rose 300–800 meters, and in May some occasional ash emissions rose as high as 5 kilometers. The 30,000 indigenous people who lived on the upper slopes of Pinatubo—the Aeta—blamed the increasing activity on geothermal exploration wells that had been drilled in the area in the late 1980s, angering their god of the mountain. Based on the steam explosions and the high number of small earthquakes recorded by the newly established net-

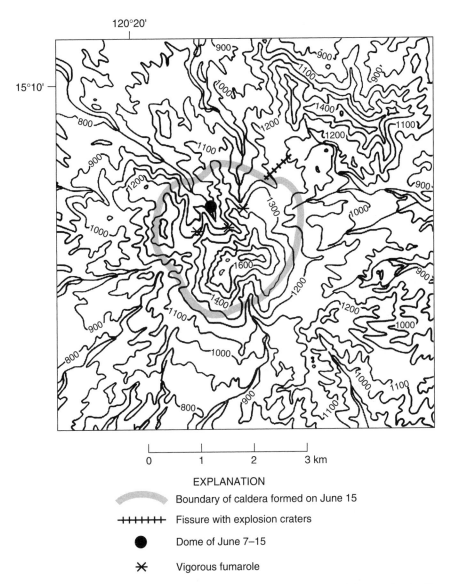

120°20'

15°10'

900

900

900

1000

1400

800

1200

1200

1100

1100

1200

900

900

1000

1200

1000

1300

1600

1200

900

1400

1200

1100

1000

900

800

1000

1000

1100

1000

800

900

1100

1100

```
L_____L_____L_____L_____L
0        1        2        3 km
```

EXPLANATION

Boundary of caldera formed on June 15

+++++++   Fissure with explosion craters

●   Dome of June 7–15

✳   Vigorous fumarole

44 Major features near the summit of Mount Pinatubo formed during the April 2–June 15, 1991, eruption. (From *Fire and Mud*, Newhall and Punongbayon, 1996, used with permission.)

work of seismometers, PHIVOLCS in early April recommended evacuation of the area within a radius of 10 kilometers of the summit; about 5000 residents left at this time.

On April 5, PHIVOLCS established the first of six seismic stations on the west side of Pinatubo. Earthquakes near and beneath a potentially active volcano provide volcanologists with some insight into what is happening inside the Earth, just as an X-ray negative provides a doctor some clue to the inner troubles of a patient. Data from this early seismic network were not "real time"; that is, the recordings had to be read and the locations of the quakes calculated from their arrival times at each seismometer. This task provided a daily earthquake count—from 33 to 178—during April and May 1991. These quakes were small events, only a few of which were felt. The 21 earthquake sources strong enough to be located in April clustered beneath an area 5 to 7 kilometers northwest of the summit of Pinatubo at depths of 2 to 7 kilometers beneath the surface (Figure 45). This was puzzling, because the earthquake swarm was neither beneath the summit nor beneath the steam-explosion fracture and active vents.

The USGS support group arrived with 35 trunks of instruments in late April. They set up their seismic receivers and office in an empty house on Clark Air Base and named it the Pinatubo Volcano Observatory (PVO). With much wangling of helicopter support they completed the installation of a seven-station seismic network around the volcano. The seismic signals were radioed into their office, and nearly real-time earthquake locations were determined by computer. Their first discovery was that PHIVOLCS's earlier earthquake locations were essentially correct. The main swarm was occurring northwest of Pinatubo's summit at a depth of about 5 kilometers below sea level. The network also provided real-time seismic amplitude measurements (RSAM), which average the seismicity received from each station over ten-minute intervals. RSAM have proved very useful in monitoring volcanic unrest. They answer the questions of whether the seismicity is increasing, and if so, by how much (Figure 46).

The seismic networks also provided a rapid way to determine the character of the earthquakes. Most of them, especially in April and May, had high-frequency signatures with a sharp onset lasting only a few seconds, as seen on the continuous recording drums. These *volcano-tectonic* (VT) quakes are interpreted as being caused by rock cracking when magma forces its way upward from underground. *Long-period* (LP) seismic events and volcanic tremor are signatures that often occur before and during eruptions. Their causes are still being debated, but most investigators believe they are related to resonant pressure variations in fluid-filled subsurface cracks.

---

45 Locations of earthquakes beneath Mount Pinatubo from May 6 to May 11, 1991: (A) epicenter map and (B) cross section. (From *Fire and Mud*, Newhall and Punongbayon, 1996, used with permission.)

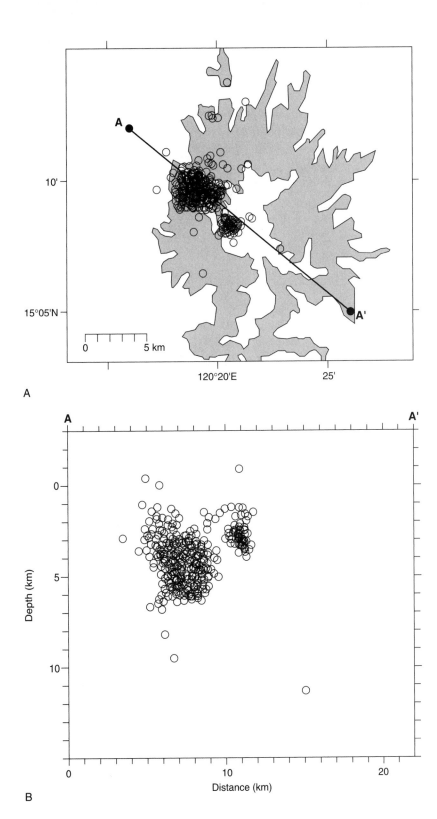

A

B

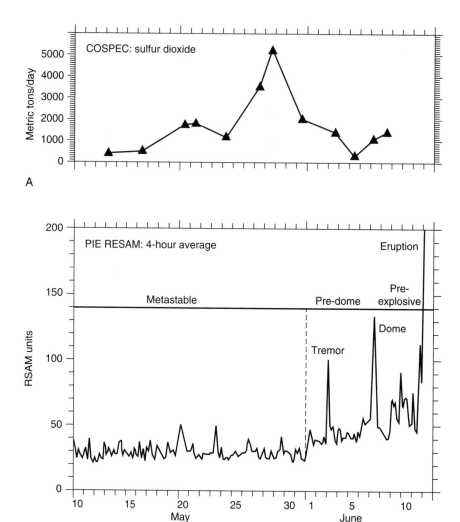

46 Records of (A) sulfur dioxide emissions and (B) real-time seismic amplitude measurements (RSAM) from Mount Pinatubo, May 10–June 12, 1991. (The RSAM are determined from the average amplitude of ground vibrations at a seismometer station averaged over ten-minute intervals. These measurements answer two important questions: How high is the seismicity, and is it increasing or decreasing?) (From *Fire and Mud*, Newhall and Punongbayon, 1996, used with permission.)

On May 12, airborne measurements of sulfur dioxide (SO_2) using a correlation spectrometer (COSPEC) were started. Emission of SO_2 increased from about 500 metric tons per day on May 13 to over 5000 tons on May 28 (see Figure 46A). However, it declined to 1400 tons per day on May 30. One interpretation of SO_2 emission is that the gas escapes from underground magma storage chambers. Its decline after May 28 was another puzzle.

Meanwhile, geologists had been examining the volcanic deposits from prehistoric eruptions of Pinatubo and collecting samples of charcoal from wood fragments charred by the intense heat of those ancient pyroclastic flows. The scale of the deposits was truly alarming—many cubic kilometers in volume, and covering large areas on the flanks of Pinatubo. The geologists were awed by the enormous pyroclastic deposits in a deep canyon near the edge of Clark Air Base, and they concluded that these once-incinerating deposits had swept across most of the area where the base was later built.

The ages of the charcoal samples were rapidly determined by radiocarbon dating at the USGS laboratory in Washington, DC. They indicated three giant explosive eruptions, or extended periods of eruptions, during the past 6000 years—one about 5500 years ago, another about 3000 years ago, and a third about A.D. 1500 (Figure 47). Was 1991 go-

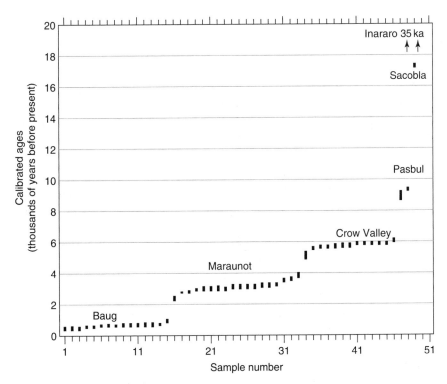

47 Radiocarbon ages of Mount Pinatubo prehistoric pyroclastic flow deposits arranged in chronologic order indicate three major periods of previous eruptions—about 500, 1500, and 3000 years ago. (From *Fire and Mud*, Newhall and Punongbayon, 1996, used with permission.)

ing to produce another huge eruption? That was the key question on the minds of the PHIVOLCS and USGS scientists during the month of May. Geology often repeats itself in a general way, seldom in a detailed way; however, even a general repeat of the prehistoric Pinatubo eruptions was an awesome prospect.

A parallel problem at that time was how to alert the hundreds of thousands of people living within the area of potential hazard seen in Pinatubo's dangerous past. If evacuations were called for, would people heed the alarm? In a relatively small volcanic eruption in Colombia in 1985, warnings of possible mudflows were largely unheeded, and 23,000 people died (see Chapter 10).

During May the Pinatubo investigative team began an intensive education program aimed at the people at risk. PHIVOLCS concentrated on the civilian population and the USGS on the Clark Air Base command and staff. It was slow going. It was nearly unthinkable to the military even to consider evacuating the base; mayors in the nearby cities were more alarmed by the potential for disruption and economic loss than by a volcanic eruption that occurred 500 years ago. Many government leaders considered the geologists to be like Chicken Little running around saying "The sky is falling."

But slowly, the educational campaign began to take hold. Air Force officers came to the PVO and learned to help the seismologists and COSPEC investigators with their data collection and equipment repairs. They were impressed that when money was not available, some of the scientists put the cost of replacement truck batteries available at the base Post Exchange on their personal credit cards. The helicopter pilots saw the volcano close up when they carried the scientists to install and repair their seismometers and radio relays and to visually monitor changes in the summit area. Working closely with the geologists and listening to their explanations of what was occurring—and might occur—went a long way in establishing their credibility.

Fortunately, PVO had copies of a video called "Understanding Volcanic Hazards" produced by Maurice and Katia Krafft, a French couple famous for their scientific investigations and films of volcanic phenomena. After people saw this video, the pictures of exploding volcanoes, pyroclastic flows, and lahars needed little additional explanation. The team showed the video to whoever would watch. It was shown on broadcast TV whenever stations would air it, and although the base command was at first reluctant, it was eventually shown on the base TV station.

Early June marked a significant change in the seismicity beneath the volcano. A new swarm of VT earthquake locations began to occur beneath the summit of Pinatubo at shallow depths, and the number of LP quakes, periods of volcanic tremor, and RSAM values increased. On June 8, a small lava dome was observed by helicopter high on the volcano, a few hundred meters north of the steam vents. This was the first evidence that new magma had reached the surface.

On June 5, PHIVOLCS issued a level 3 alert (major pyroclastic eruption possible within 2 weeks) based on the increased seismicity. On June 7, PHIVOLCS increased the alert to level 4 (eruption possible within 24 hours), based on seismicity and on summit inflation seen on a recently installed electronic tiltmeter near one of the upper-slope seismometers. This upgrade caused considerable confusion. The USGS group at Clark was still at level 3, and the Air Force, although it had an evacuation plan for the base, was still not ready to put the plan into action. Seismicity waxed and waned on a hourly to daily basis, and so did the anxiety of the observers. On June 9 the Air Force general and his base commander helicoptered over Pinatubo with one of the PVO geologists. They saw the dome and the increased ash emissions, and their proximity to the base. The general turned to his base commander and said "Do it tomorrow." The evacuation decision had been made.

At 6:00 A.M. on June 10, the base evacuation began as planned. Vehicles laden with staff and dependents poured out of the gate in military precision, bound for the Subic Bay Naval Base, 40 kilometers south of Pinatubo. By noon, nearly 15,000 people had left. All remaining aircraft except three helicopters also departed. This impressive evacuation helped PHIVOLCS convince the much larger population of farmers and city dwellers who lived on the fertile lower slopes of Pinatubo that the U.S. Air Force took the threat very seriously. Now what? PVO moved most of its staff 3 miles east to a building just inside the base wall. About 1200 base personnel stayed to provide security. RSAM and LP quakes continued to increase. The wait was agonizing.

On a clear morning on June 12, the wait was over. A huge explosive eruption column soared upward (Plate 16). It reached 19 kilometers in height, and its top mushroomed into a dense shape frighteningly similar to a nuclear mushroom cloud, but without an audible sound. By afternoon the explosion column had waned, and the RSAM had dropped back to June 11 levels. But seeing was believing. Base security was reduced to 600 soldiers, PVO completed its move to a more distant building on the base, the general evacuation radius was extended to 30 kilometers, and the total number of evacuees increased to about 60,000. Was the worst over?

Not yet. That evening the seismometers began jiggling wildly again. Another eruption column topped 25 kilometers on the base radar. Lightning from the huge buildup of static electricity in the volcanic ash clouds

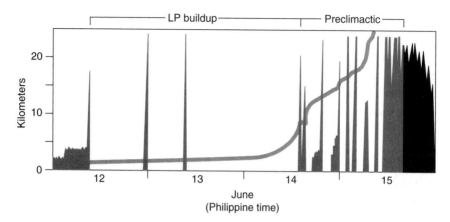

48 The buildup to the climactic eruption of Mount Pinatubo from June 12 through June 15, 1991, as shown by the altitudes of the explosion clouds (vertical bars), and the RSAM (solid gray line). (From *Fire and Mud*, Newhall and Punongbayon, 1996, used with permission.)

lit the night sky. After a pause, another major explosive eruption occurred on the morning of the June 13. To complicate matters, a typhoon with 160-kilometer-per-hour winds was approaching the base. On the afternoon of June 14, the volcano erupted its first big series of ground-hugging pyroclastic flows (Figure 48).

At dawn on June 15, the storm and the eruption climax began their deadly convergence. Within minutes, the roiling ash cloud was more than 7 kilometers wide, its top hidden in the typhoon clouds. The eruption built in huge pulses, and finally, at 1:42 P.M., the giant, continuous climactic eruption began (Figure 49). The pulses were hidden from the base observers; their timing was reconstructed from barograph records rattled by the low-frequency sound waves from the disgorging volcano. Satellite photos showed that the ash cloud reached a maximum height of 34 kilometers, punching far above the typhoon rain clouds.

Back at Clark, heavy rain and ash with pumice lumps the size of golf balls pummeled the PVO building. Only one seismometer was still operating, and its continuous tremor signal was off the scale. The only thing left to do was leave and hope to survive. The geologists, the Air Force officers, and the remaining MPs left in the darkness in a caravan of cars separated by only a few feet so that the tail lights of the car ahead could still be seen. Rain, a sludge of gray ash, and pumice lumps impeded the caravan heading to its last-ditch refuge 34 kilometers from Pinatubo. Was it far enough away? Prehistoric pyroclastic flows from giant calderas such as Crater Lake were known to have traveled farther.

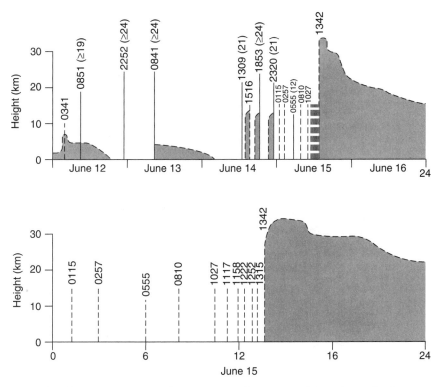

49 The climactic eruption of Mount Pinatubo on June 15, as shown by the times and altitudes of the individual explosion clouds and the continuous eruption column that began at 1:42 P.M. (From *Fire and Mud*, Newhall and Punongbayon, 1996, used with permission.)

As one of the geologists recalled, "it was like running away in slow motion. . . . We're going maybe 20 miles per hour. I'd rather have been going eighty, but we couldn't go faster. We couldn't see, and there was lots of traffic." Three hours later, the caravan reached the new headquarters 16 kilometers northeast of Clark. Big quakes jolted the building, but it held, and the metal roof did not collapse from the heavy load of wet ash. The volcano boiled on for several more hours before winding down. The worst was finally over (Figure 50).

The eye of Typhoon Yunya passed some 100 kilometers northeast of Pinatubo at about 11:00 A.M. on June 15, as the volcano was building to its final huge eruption. The storm's heavy rains, however, continued. The changing wind patterns distributed the heavy, rain-soaked ash over a large area surrounding the volcano (Figure 51). Roofs collapsing from the load killed most of the 250 victims of the eruption. The torrential rains also triggered massive lahars, which swept up the newly deposited ash and py-

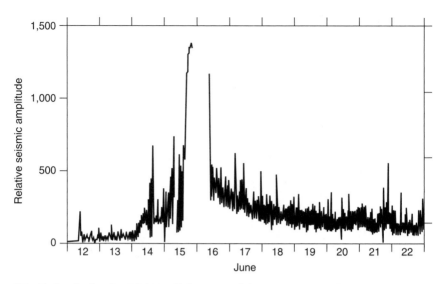

50 In hindsight, the RSAM tell the story of the Mount Pinatubo eruption. From June 12 to 15, even though pauses in the seismicity led to uncertainty about the next eruptive event, the trend of the buildup to the cataclysmic eruption is clear. On June 15 and 16, seismic energy defines the climax eruption. From June 16 to 22, the exponential trend of decreasing seismicity indicates that the eruption is waning. (From *Fire and Mud*, Newhall and Punongbayon, 1996, used with permission.)

roclastic flow deposits into thick floods of mud and debris that churned down all sides of the volcano. Some of the lahars that eroded the thick, newly formed pyroclastic flow deposits were scalding hot.

The curse of lahars is that they do not end when a volcanic eruption stops. In places like the Philippines, with a yearly rainy season, lahars continue for many years, until the load of loose volcanic debris has been redistributed down the volcano's flanks onto the alluvial aprons surrounding it. At Pinatubo, those aprons were prime farmland and, along with their supporting cities and towns, accounted for the large population surrounding the volcano. Most of the economic losses caused by the eruption of Pinatubo resulted from the destruction of homes and farmlands as the lahar floods overtopped their stream channels and buried the floodplains with thick deposits of mud, sand, and gravel. Over centuries, these mudflow deposits will once again provide fertile farmland. However, the process of soil recovery is slow, and the needs of a displaced farm family are immediate.

In summary, the eruption of Pinatubo spewed out an estimated 3 to 4 cubic kilometers of airfall ash and pumice fragments and 5 to 6 cubic

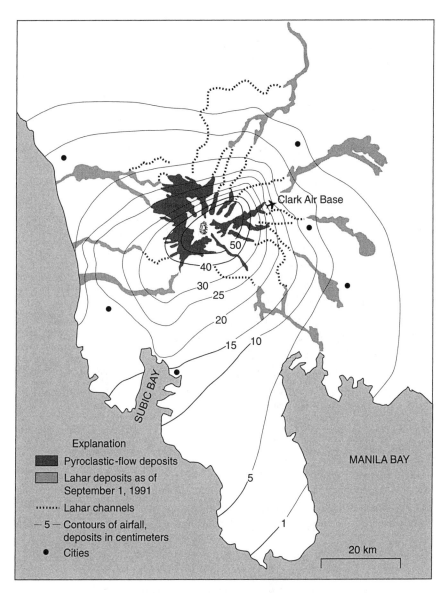

51 Map showing the major deposits from the eruption of Mount Pinatubo in 1991. The airfall ash deposits were from the explosions of June 12 to 15, the pyroclastic flow deposits were mainly from the climactic eruption on June 15, and lahars continued to plague the region every rainy season for years after the eruption. (After the Pinatubo Volcano Observatory Team, *Eos*, December 3, 1991.)

kilometers of pyroclastic flow deposits that extended as far as 16 kilometers from the old summit. The volume of magma erupted was about 5 cubic kilometers, and the old summit was replaced by a 2.5-kilometer-wide caldera. The high point on the caldera rim was 1486 meters, 259 meters below the pre-eruption summit. Small ash eruptions continued until September 1991, and a lava dome formed within the caldera in 1992.

By the end of 1991, 300 square kilometers of land surrounding Pinatubo had been covered by the lahars. Their occurrence during the monsoon season in the years following the eruption has decreased, but they remained destructive for a decade after the great eruption. The rains also created a caldera lake. Its surface elevation in 1991 was 780 meters, and by 2001 it had reached 955 meters, only 5 meters below spilling over. To avoid the possibility of sudden flooding from failure of the caldera rim, a spillway was constructed to stabilize the lake.

The June 15, 1991, eruption of Mount Pinatubo was the second largest in the twentieth century, exceeded in volume of magma only by the Katmai eruption in Alaska. The enormous amounts of ash and aerosols injected into the stratosphere by the Pinatubo eruption had a profound effect on the global climate (see Chapter 17). That's the geologic story; the courage and perseverance of the hundreds of thousands of Philippine people who endured and survived this huge volcanic eruption is the great human story of Pinatubo.

Links

MTU Volcanoes Page—Pinatubo
  http://www.geo.mtu.edu/volcanoes/pinatubo/volcano/

NOAA—Pinatubo Teachers Guide
  http://www.ngdc.noaa.gov/seg/hazard/stratoguide/pinfact.html

Pinatubo Volcano
  http://www.park.org/Philippines/pinatubo/

Smithsonian Global Volcanism Program—Pinatubo
  http://www.volcano.si.edu/gvp/world/volcano.cfm?vnum=0703-083

U.S. Geological Survey Cascade Volcano Observatory—Pinatubo
  http://vulcan.wr.usgs.gov/Volcanoes/Philippines/Pinatubo/framework.html

U.S. Geological Survey, Fire and Mud: Eruptions and Lahars of Mount Pinatubo, Philippines, 1996
  http://pubs.usgs.gov/pinatubo/

Volcano World—Pinatubo
  http://volcano.und.nodak.edu/vwdocs/volc_images/southeast_asia/philippines/pinatubo.html

Sources

Endo, E. T., and T. Murray. "Real-Time Seismic Amplitude Measurement (RSAM): A Volcano Monitoring and Prediction tool." *Bulletin of Volcanology*, 53 (1991): 533–545.

Krafft, Maurice, and Katia Krafft. (Video.) *Understanding Volcanic Hazards*. Available from Northwest Interpretive Association, 900 First Ave., Suite 9, Seattle, WA 98104.

Lee, W. H. K., ed. *Toolbox for Seismic Data Acquisition, Processing, and Analysis*. El Cerrito, CA: International Association of Seismology and Physics of the Earth's Interior in collaboration with Seismological Society of America, 1989. 283 pp.

Newhall, C. G., and R. S. Punongbayon, eds. *Fire and Mud: Eruptions and Lahars of Mount Pinatubo, Philippines*. Seattle: University of Washington Press, 1996.

Pallister, J. S., R. P. Hoblitt, and A. G. Reyes. "A Basalt Trigger for the 1991 Eruptions of Pinatubo Volcano?" *Nature* 356 (1992): 426–428.

CHAPTER 6

Ring of Fire

52   Cerro Negro Volcano, Nicaragua, 1968.

America is drifting away from Europe and toward the Philippine Islands. In fact, all the plates around the Pacific are slowly converging along the subduction zones that encircle this vast ocean. Volcanoes and earthquakes along these convergent plate margins form a "Ring of Fire" that nearly surrounds the Pacific basin (Figure 53).

Mount St. Helens and the other volcanoes of the Cascade Range are located on the subduction zone between the small Juan de Fuca Plate and the North America Plate. Volcanoes in Mexico, Central America, the Andes, Antarctica, New Zealand, Papua New Guinea, the Philippines, Japan, Kamchatka, and Alaska complete the circuit of the Ring of Fire. Another major belt of convergent plate margins runs from Indonesia through the Himalayas and the Middle East to Greece and Italy. Where continental edges of two plates converge, as in the Himalayas, volcanoes are rare and the earthquake zones more diffuse.

Subduction zones generate about 400 of the world's more than 500 known *active* volcanoes; that is, those that have erupted at least once in recorded history. The Mediterranean volcanoes of Italy and Greece are included in this tally.

There are major differences between subduction volcanoes and the volcanoes along oceanic rifts. *Subduction volcanoes* form island arcs and high mountain chains rather than submarine ridges; they are more explosive and produce large volumes of ash as well as lava flows; and their products are more variable in composition. Even their basic shapes as individual mountains differ from those of volcanoes that form at divergent plate margins. *Rift volcanoes* are located at the exact edges of the separating plates, while subduction volcanoes occur on the overriding plate about 100 to 200 kilometers landward from the deep-ocean trenches that mark the edges of the converging plates.

Chains of subduction volcanoes form graceful arcs, thousands of kilometers in length, across the globe. A close look often shows five or ten volcanoes in a fairly straight line a few hundred kilometers in length, with a slight bend, or *offset*, between adjoining groups. The offsets between the groups form the overall arc, which is generally convex toward the open ocean side.

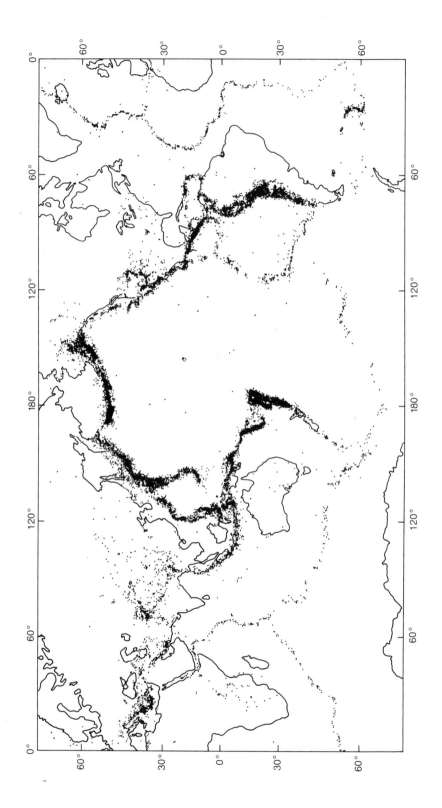

Plate 1 High lava fountains and rivers of molten rock repeatedly gush out during the episodic phase of Kilauea Volcano's Pu'u O'o eruption. (Photograph by J. D. Griggs, U.S. Geological Survey, January 31, 1984.)

Plate 2 A river of molten lava 10 to 20 meters wide flows from Pu'u O'o and forms the central channel of a black a'a lava flow. The trees on the left side of the flow are about 15 meters high. (Photograph by J. D. Griggs, U.S. Geological Survey.)

Plate 3   An arching lava fountain, 10 to 15 meters high, is hurled from an early vent of the Pu'u O'o eruption. (Photograph by J. D. Griggs, U.S. Geological Survey.)

Plate 4   Lava from the Pu'u O'o eruption of Kilauea Volcano, Hawaii, travels in shallow underground tubes for 6 miles toward the sea. Sometimes lava breaks out of its tube, as in this photo, and forms a large glowing mass several meters wide. (Photograph by Don Swanson, U.S. Geological Survey, 2002.)

Plate 5 A 400-meter-high lava fountain during episode 31 of the Pu'u O'o eruption dwarfs a civil defense helicopter. (Photograph by J. D. Griggs, U.S. Geological Survey.)

Plate 6 The main lava flow from Mauna Loa's 1984 eruption poured from a vent at 2900 meters elevation and traveled 25 kilometers down the mountain in five days to an elevation of 900 meters, only 6 kilometers from the city of Hilo. (Photograph by David Little.)

Plate 7 A pahoehoe lava flow from Kilauea Volcano's Pu'u O'o eruption creeps across a lawn before destroying another home on it way to the sea. (Photograph by J. D. Griggs, U.S. Geological Survey.)

Plate 8 Lava from the Pu'u O'o eruption pours over a high cliff into the sea. (Photograph by Don Swanson, U.S. Geological Survey, 2002.)

Plate 9   While photographing lava streams flowing down Mauna Loa Volcano during its 1984 eruption, Katia Krafft captured this dramatic moment, which she named "Pele Dancing" to honor the Hawaiian goddess of volcanoes.

Plate 10 An eruption of the Soufrière Hills Volcano on Montserrat, an island in the Caribbean Sea, began in 1995 and has continued through 2004. A growing lava dome has collapsed and exploded several times, building a delta of avalanche and pyroclastic flow deposits. (Copyright photo by Steve O'Meara.)

53 The Ring of Fire around the Pacific Ocean, delineated by the locations of earthquakes occurring over a decade with a magnitude greater than 4.5, as determined by the U.S. Geological Survey's National Earthquake Information Service. (Map plotted by the Environmental Data and Information Service of the National Oceanic and Atmospheric Administration.)

◄───

Japan is a good example of a subduction chain, with about 50 volcanoes along parts of four island arcs. The deep-sea trenches lie 200 kilometers to the Pacific Ocean side of the volcanoes and mark the actual boundary between the plates. A zone of earthquakes dips from near the Earth's surface at the trenches to below the island arcs. This zone roughly outlines the edges of the Pacific and Philippine plates as they plunge beneath Asia.

The belt of volcanoes grows on the overlying plate, where the earthquake zone is 100 to 200 kilometers beneath the Earth's surface (Figure 54). Most of the large volcanoes are on the eastern edge of this belt, where the earthquake zone is more shallow. Japanese geologists refer to this eastern edge as the *volcanic front*. The number of volcanoes and their production of volcanic rocks decreases gradually with distance away from the volcanic front toward Asia, but stops abruptly on the Pacific side of the front.

This pattern is generally characteristic of subduction-zone volcanoes and indicates that magma is generated when the descending plate sinks to a depth of 100–200 kilometers. At one time geologists believed that

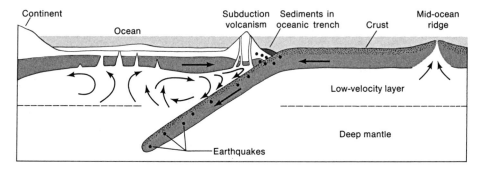

54 Formation of a belt of subduction volcanoes. The plate of crust and solid upper mantle formed at a mid-ocean ridge moves over the plastic low-velocity layer and dives beneath a converging plate at a subduction zone. An oceanic trench forms where the plates converge, and an island arc of compressed rocks and volcanoes is built over the sinking plate. (From M. Nafi Toksoz, "The Subduction of the Lithosphere." Copyright © 1975 by Scientific American, Inc. All rights reserved.)

the friction of the converging plates generated heat and that this extra heat helped melt the magma for subduction volcanoes. Numerical models of the temperatures beneath subduction zones show, however, that the underthrusting plate carries its lower surface temperature down with it, in effect cooling part of the region deep beneath the volcanic front. Most geologists now believe that the magma generated in subduction zones is formed by the fluxing action of water and calcium carbonate in the oceanic sediments dragged down on the top of the underthrusting plate. Water, carbon dioxide, and minerals that contain sodium and potassium lower the melting temperature of rock. The existing temperatures in the low-velocity layer at a depth of 100–200 kilometers along the top of the subducting plate are high enough to melt this combination of rock and volatiles. The addition of water and calcium carbonate also increases the amount of gas that boils out of magma as it rises toward the surface, and it explains the greater explosiveness of subduction volcanoes.

When enough magma has formed at depth, its lesser density exerts an upward force, and its ascent toward the surface may be aided by fractures in the overlying plate. The distance between volcanoes in island arcs is usually about 50 to 70 kilometers, a spacing that is about the same as the thickness of the overlying plate. In experimental models, the distance between fractures created in a rigid plate resting on a soft layer tends to be about the same as the thickness of the plate.

Figure 55 shows a good conceptual model of the island of Sumatra in Indonesia. The Australia Plate on the left is plunging beneath the continental portion of the Eurasia Plate on the right. Scrapings from the seafloor of the Indian Ocean pile up into a submarine *accretionary wedge* and islands of deformed sediments called the *outer-arc high*. The back-arc region to the right of the active volcanoes in Sumatra is filled with oil-rich sedimentary rocks.

In some places, the subducted plate dips at a low angle; in others, at a nearly vertical angle. Christopher Scholz, a geophysicist at Columbia University, and Jaime Campos, a seismologist at the University of Chile, proposed a novel explanation for these variations. In their "sea anchor" model, the dip of the subducted slab—and therefore the nature of the volcanoes and structures on the overlying slab—depends on the relative movement of the two plates and the movement of the underlying mantle. Where the plates are pushing toward each other, as in the Andes, the dip is shallow. Where both the plates are moving in the same direction, as in the Izu–Bonin Islands south of Japan, the dip is about 45°. And where the subducting plate is moving in the same direction as the overlying plate at nearly the same speed, as in the Mariana Islands in the

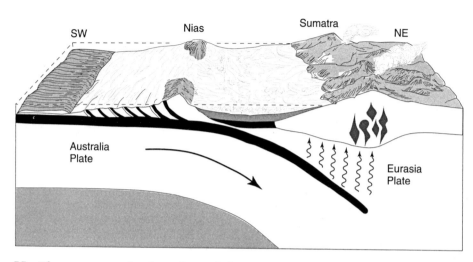

55 The convergent plate boundary in Indonesia, where the Australia Plate is subducted beneath the Eurasia Plate. The outer-arc high forms the chain of small islands that includes Nias, made up of sedimentary rocks. The subduction volcanoes of Sumatra, on the inner island arc, form above where the surface of the subducted plate reaches a depth of 100–150 kilometers. Water and carbon dioxide derived from the sediments carried down on the Australia Plate reduce the melting temperature in the upper mantle beneath Sumatra and contribute to the formation of magma. Bodies of magma then rise because they are lighter than the surrounding, unmelted rock. Some of the magma solidifies within the crust, and some escapes to the surface to form the active volcanoes. (After B. Clark Burchfiel, "The Continental Crust." Copyright © 1983 by Scientific American, Inc. All rights reserved.

Western Pacific, the dip is nearly vertical (Figure 56). Scholz and Campos compare these situations to the effect of the ocean current versus the wind direction on the angle of a sea anchor cast from a boat. This is an attractive, albeit simple, model, and in a complex world simplicity is most welcome.

The silica content of most rift volcanoes is about 50 percent, varying by only a few percentage points, but the silica content of subduction volcanoes varies from about 50 to 75 percent. The silica content largely determines which minerals crystallize from the cooling melt, and thereby controls the type of volcanic rocks that are formed and the sediments that are derived from their erosion. For example, the mineral *quartz* (SiO_2) does not form during crystallization until the silica content of magma exceeds about 55 percent. In Iceland there is no quartz in the basaltic lavas, and there are no white, sandy beaches composed of quartz grains. On the Ring of Fire, however, the more silica-rich volcanic rocks and the

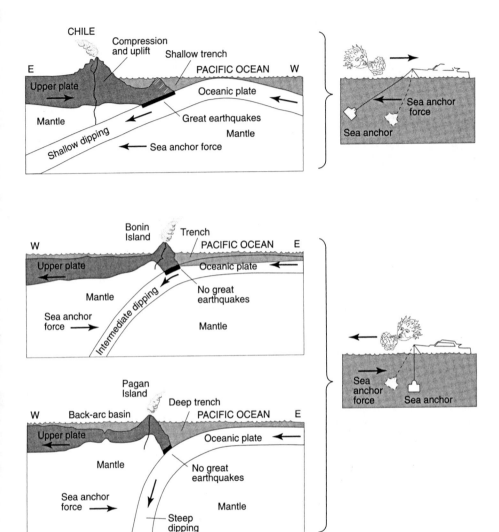

56 Christopher Scholz at Columbia University and Jaime Campos at the University of Chile developed this cartoon to explain variations in the dip and character of subduction zones. Their concept of a "sea anchor force" compares the motion of a boat relative to that of the sea, and the resulting angle of a rope attached to a sea anchor, with the motion of the upper plate in a subduction zone (analogous to the wind) relative to that of the Earth's mantle (analogous to the ocean current). In the top example, the Nazca Plate (oceanic) moves east, and the South America Plate (Chile) moves west. The sea anchor force pushes east with the descending slab, resulting in a shallow-dipping subduction zone. This in turn causes a wide area of high friction between the two plates, large earthquakes, and a wide separation between the oceanic trench and the belt of volcanoes. In the middle example, south of Japan, both the Philippine Plate (upper) and the Pacific Plate (oceanic) are moving west. The sea anchor force pushes against the granitic rocks beneath them often contain quartz grains that erode to form white, quartz-sand beaches.

The composition of volcanic rocks is principally controlled by three factors: partial melting, partial crystallization, and contamination by surrounding rocks. *Partial melting* is the process by which part of the solid upper mantle or lower crust of the Earth is turned into magma. Rocks are a mixture of several chemical compounds, and not all of them have the same melting points. In partial melting, the first fraction to melt will have a larger proportion of the low-melting-point compounds than the parent rock. An analogy is the extraction of alcohol from hard cider by freezing and thawing. If a completely frozen block of hard cider is slowly melted, the first fraction to melt will contain a higher proportion of alcohol than does the remaining solid block.

Partial crystallization is just the opposite. As molten rock slowly cools, the minerals with higher melting points crystallize first, and the residual melt keeps changing in composition. This process is analogous to slowly freezing the hard cider and thereby concentrating the alcohol in the residual unfrozen liquid.

Silica is concentrated in the low-melting-point fractions, especially when water is present. Therefore, both partial melting with small quantities of water present and partial crystallization during the ascent of magma to the surface produce a liquid with a higher silica content than that of the starting material. In general, then, magmas are richer in silica than are their upper-mantle sources and may be further "distilled" during crystallization to produce an even more silica-rich rock.

Magma rises greater distances in subduction volcanoes than it does in rift volcanoes. Furthermore, the magma that supplies subduction volcanoes often must ascend through ancient continental rocks, already high in silica, which greatly increases their chance of *contamination*.

---

subducting plate, causing it to dip more steeply. The resulting friction between the plates is less; the number of large earthquakes is smaller; and the separation between the trench and the volcanic belt is smaller. In the bottom example, across the Northern Mariana Islands, the rearward part of the Philippine Plate (upper) is moving at nearly the same speed as the Pacific Plate (oceanic). The increased sea anchor force pushes the subducting slab to a nearly vertical dip, and the distance between the trench and the volcanic belt is small. A back-arc basin forms west of the subduction volcanoes to accommodate the difference in motion between the island arc and trench and the rearward part of the Philippine Plate. (After C. H. Scholz and J. Campos, "On the Mechanism of Seismic Decoupling and Back Arc Spreading at Subduction Zones," *Journal of Geophysical Research* 100 [1995]: 103–115; and *Eos*, December 5, 1995.)

These concepts of partial melting, partial crystallization, and contamination provide general explanations of the origin of the many diverse volcanic rocks. Several more specific explanations have also been proposed, and the arguments favoring one process over another wax and wane on subtle details. The problem is not a lack of explanations, but too many explanations. In fact, more than one process of variation is probably at work. Truth is often a mixture of conflicting opinions.

The greater proportion of water and carbon dioxide dissolved in the magma of subduction volcanoes, coupled with slightly lower eruption temperatures than in rift volcanoes, leads to more explosive eruptions. The gas wants out, and the cooler and therefore more viscous magma is holding it back, resulting in explosive eruptions of volcanic ash. Subduction eruptions often begin with explosive showers of ash and lumps of pumice and end with thick, viscous lava flows. Such alternations of volcanic ash and lava flows build beautiful steep cones like that of Mount Fuji, the archetype of classic volcanic form (Figure 57).

57 Mount Fuji, or Fuji-san, the archetypal volcano. The classic shape of this nearly perfect cone with its graceful concave slopes—and its location near Tokyo—make it the world's best-known volcano. It rises to 3776 meters from a base 30 kilometers in diameter almost at sea level. Fuji-san embodies the beauty, majesty, and power of all nature. (Woodblock print by Hokusai [1760–1849] from his famous series *Thirty-Six Views of Mount Fuji.*)

About 40 subduction volcanoes erupt each year. A few, like Stromboli in the Mediterranean, have been erupting nearly continuously for centuries. Others, like Mount Pinatubo, have erupted only once in recorded history. In general, the longer the period between eruptions, the greater the likelihood that the next eruption will be a big one.

The products of subduction volcanoes are largely explosive: ash, pumice, cinders, blocks, and molten lava bombs. This is in sharp contrast to the rift volcanoes along the oceanic ridges, whose products are mainly effusive (nonexplosive) lava flows. The more notorious of the world's volcanoes belong to the subduction clan: Mount St. Helens and Mount Pinatubo, as described in the previous two chapters, as well as Krakatau, Vesuvius, Mont Pelée, Katmai, Bezymianny, El Chichón, Montserrat, and Nevado del Ruiz.

Vesuvius buried Pompeii in A.D. 79 and provided a time capsule of Roman art, architecture, and artifacts for its nineteenth- and twentieth-century excavators (Figure 58). Before its violent eruption in Roman

58 Pompeii was buried beneath 6 meters of volcanic ash for nearly 2000 years. Its destroyer and preserver, Mount Vesuvius, can be seen in the background. Airfall ash and pyroclastic flows buried the town in one day in A.D. 79. Pliny, a Roman historian, described the eruption cloud as looking like a pine tree—a puzzling comparison until you note the tree behind the ruins.

times, Vesuvius was considered extinct, but it has erupted many times since then, most recently in 1944.

Mont Pelée, on the island of Martinique in the West Indies, annihilated the 28,000 inhabitants of the port city of St. Pierre on May 8, 1902. A pyroclastic flow—what French geologists call a *nuée ardente* (glowing cloud)—rushed down the mountainside, and the hot gases and ash bowled over and burned everything in their path, including ships in the harbor. The disaster lasted only a few minutes and left just two survivors in St. Pierre.

Katmai, on the Alaska Peninsula, erupted for two days in June 1912. The ashfall totaled 20 cubic kilometers and nearly buried Kodiak Island. The summit of Mount Katmai sank into a caldera 5 kilometers wide, and 10 cubic kilometers of pyroclastic flow deposits filled a valley 20 kilometers long to form the Valley of Ten Thousand Smokes. The eruption of Katmai was the largest in the twentieth century.

Bezymianny, on the Kamchatka peninsula of Siberia, erupted for the first time in recorded history on October 22, 1955, culminating in a giant explosive eruption on March 30, 1956 (Figure 59). The main eruption cloud reached 35 kilometers in height; large rocks were hurled 25 kilometers away; a caldera 2 kilometers wide was formed; and pyroclas-

59 The giant explosion cloud from Bezymianny Volcano, Kamchatka, March 30, 1956. The lower edge of the huge, fan-shaped ash cloud is about 7 kilometers high, and the top is about 35 kilometers high. (Photograph by I. V. Erov from 45 kilometers west of the volcano.)

tic flows dumped 3 cubic kilometers of searing hot deposits that Russian geologists call the Valley of Ten Thousand Smokes of Kamchatka.

Fortunately, the areas surrounding both Katmai and Bezymianny were almost uninhabited, and no one was killed. Such eruptions in Japan or Central America—or anywhere on the Ring of Fire that is densely populated—could be catastrophic. A case in point is the giant explosive eruption of Mount Pinatubo in the Philippines in 1991. More than 500,000 people lived in the surrounding area, but fortunately most of them were able to evacuate before the climactic eruption, as described in Chapter 5. But a volcanic eruption does not have to be large to cause a catastrophe. Mudflows from the relatively small eruption of Nevado del Ruiz Volcano in Colombia in 1985 killed more than 20,000 people, a tragic story that is told in Chapter 10.

In prehistoric times, extremely large eruptions deposited enormous pyroclastic flows of pumice and ash covering thousands of square kilometers at several places around the Ring of Fire. Volumes of magma more than 200 times greater than that hurled from Pinatubo were spewed out in single eruptions. The message of such gigantic eruptions is not imminent doom; they do not occur that often. To us their message is attention and respect; since the millions of people who reside on the Ring of Fire must live with these volcanic eruptions, it makes sense to try to understand them better.

Links

Dynamic Earth—Subduction Zones
http://earth.leeds.ac.uk/dynamicearth/subduction/

Juan de Fuca Plate Subduction
http://vulcan.wr.usgs.gov/Glossary/PlateTectonics/Maps/map_juan_de_fuca_subduction.html

Subduction (Volcano World Website)
http://volcano.und.nodak.edu/vwdocs/vwlessons/plate_tectonics/part10.html

Subduction Zone Studies
http://www.ruf.rice.edu/~leeman/

State of the Arc Conference, 2003
http://www.ruf.rice.edu/~leeman/SOTA2003/conf_report.html

U.S. Geological Survey Alaska Volcanoes List
http://geopubs.wr.usgs.gov/fact-sheet/fs118-00/

Volcanic and Magmatic Studies Group
http://www.vmsg.org.uk/

Washington University in St. Louis MARGINS Program
http://www.margins.wustl.edu/Home.html

MARGINS Subduction Factory
http://www.margins.wustl.edu/SF/SF.html

Sources

Bebout, G. E., D. W. Scholl, S. H. Kirby, and J. P. Platt, eds. *Subduction: Top to Bottom*. Geophysical Monograph 96. Washington, DC: American Geophysical Union, 1996.

Eiler, John, ed. *Inside the Subduction Factory*. Geophysical Monograph 138. Washington, DC: American Geophysical Union, 2004.

Elliott, T., T. Plank, A. Zindler, W. White, and B. Bourdon, Element transport from slab to volcanic front at the Mariana arc. *Journal of Geophysical Research* 102 (1997): 14991–15019.

Hawkesworth, C. J., K. Gallagher, J. M. Hergt, and F. McDermott. "Mantle and Slab Contribution in Arc Magmas." *Annual Review of Earth and Planetary Sciences* 21 (1993): 175–204.

Simkin, Tom, and Lee Siebert. *Volcanoes of the World*. 2nd ed. Washington, DC: Smithsonian Institution Press, 1994.

CHAPTER 7

Kilauea, Hawaii

60 Kilauea Iki, 1959. (Photograph by J. P. Eaton, U.S. Geological Survey.)

Kilauea Volcano on the island of Hawaii, also known as the Big Island, is the most thoroughly studied volcano in the world (Figures 61 and 62). It is also one of the most active, with frequent eruptions and glowing lava lakes that stir and boil for years at a time. Almost all of Kilauea's eruptions are relatively quiet outpourings of fluid lavas. The word "quiet" can be misleading, though, for the vents often spray roaring fountains of incandescent lava several hundred meters into the air (Figure 60), which falls back into lava ponds or feeds lava flows. These outpourings of lava are quiet only in contrast to explosive volcanic eruptions, which produce fragmental debris and look more like huge detonations of dynamite.

Effusive eruptions of the Hawaiian type are comparatively safe to study at close range, and scientists at the Hawaiian Volcano Observatory have been at it since 1912. We were fortunate to be studying volcanoes in Hawaii during the early 1980s when major eruptions of both Kilauea and Mauna Loa volcanoes took place.

The new year of 1983 saw the start of the longest and largest rift eruption of Kilauea since written records began in 1823. By the start of 2004, the eruption, which showed no sign of stopping soon, had poured forth 2600 million cubic meters of lava (2.6 cubic kilometers)—enough to build a pyramid 1.75 kilometers (1.1 mile) high. These lava flows had traveled 11 kilometers to the sea, covering 117 square kilometers of land

---

61   The island of Hawaii, also known as the Big Island, is built of five volcanoes. The elevations in map A are in meters. The rift zones shown in gray on map B are zones of weakness from which numerous flank eruptions have originated. Kilauea is erupting at the time of this writing (2004); Mauna Loa last erupted in 1984 and Hualalai in 1801. Mauna Kea's latest eruptions were about 4000 years ago, and Kohala's about 60,000 years ago. Loihi is a submarine volcano. It is presumed to be active because recurring earthquake swarms have been recorded beneath it and because hot springs and young lava have been observed on Loihi from submersibles. (After D. W. Peterson and R. B. Moore, U.S. Geological Survey Professional Paper 1350, 1987, p. 151.)

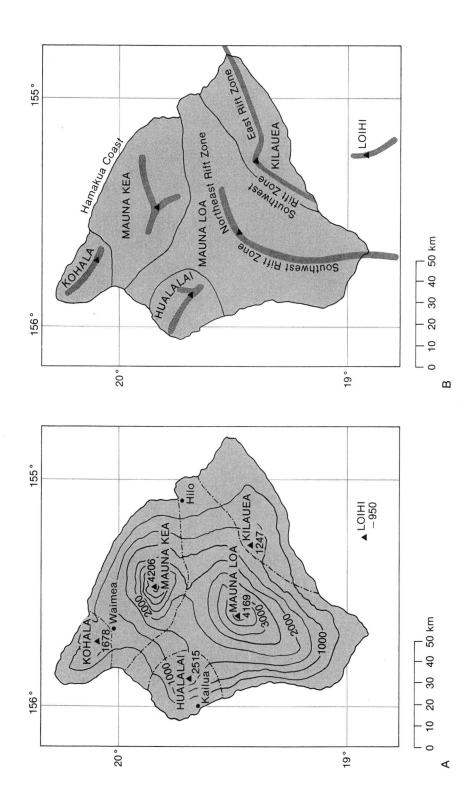

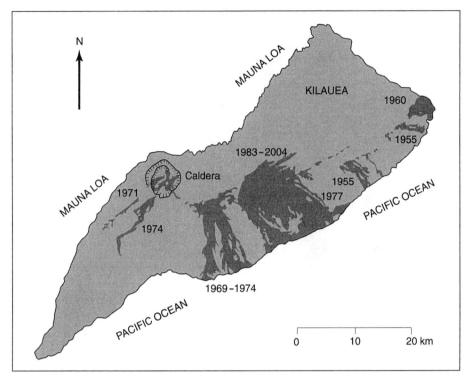

62 Most of the lava erupted from Kilauea since 1955 has come from the East Rift Zone. Two eruptions (1971 and 1974) took place on the Southwest Rift Zone and twelve in the summit caldera area. The flows labeled 1983–2004 are from the Pu'u O'o eruption. (Robin Holcomb and Christina Heliker, U.S. Geological Survey.)

and adding 2.3 square kilometers of new land to the island. Four people had been killed in accidents related to their failure to heed the posted warnings not to get too close to where the lava entered the sea; 189 buildings, including a National Park Visitor Center, had been destroyed; 14 kilometers (9 miles) of highway had been covered by lava flows; and damage estimates exceeded $60 million.

January 1, 1983, was calm, but that was soon to change. Not long after midnight, the tremor alarm triggered phone calls to staff members of the Hawaiian Volcano Observatory. Hurrying to the observatory, which is perched on the rim of Kilauea's 3-by-5-kilometer-wide caldera, they soon found all three signs (volcanic tremor, earthquake swarms, and summit deflation) of an impending eruption.

Volcanic tremor, a continuous low-frequency vibration of the ground, was being recorded on seismographs near the East Rift Zone of Kilauea.

Volcanic tremor occurs when magma is moving rapidly through underground conduits. Seismographs also showed *earthquake swarms* at shallow depths beneath the east rift, indicating that new fractures were forming, and tiltmeters registered *summit deflation*. (A *tiltmeter* works like an extremely sensitive carpenter's level and can detect changes in slope of 1 part per million—equivalent to a 1-millimeter rise of the end of an imaginary level that is 1 kilometer long.)

The volcanic tremor, earthquake swarms, and summit deflation all indicated that magma from a storage zone beneath Kilauea Caldera was being injected into the east rift (Figures 63 and 64). These shallow intrusions into the weak rift zones on the flank of the volcano are not uncommon; they may happen several times a year. Some reach the surface and erupt; others do not. The same question was on everyone's mind: would this be another aborted alarm, or would this shallow intrusion become an eruption?

All day on January 2 the tremor, earthquakes, and subsidence continued. The earthquakes were clustered on the middle east rift, about 10 to 20 kilometers east of Kilauea's summit. Finally, after midnight on January 3, lava fountains broke out in Napau Crater, 15 kilometers from the summit on the East Rift Zone. By morning, fissures in a broken line 6 kilometers long were erupting in a "curtain of fire" in a remote, inaccessible forest. During the next 12 days, various vents along this fissure system erupted intermittently, but only geologists and TV crews in helicopters were able to see the spectacular start of this long-lasting eruption.

There had been a lengthy prelude to this outbreak of lava, and many clues had come from careful observations of small earthquakes and deformation of the ground surface. Kilauea's magma rises from a depth of 60 kilometers or more and accumulates temporarily in a reservoir a few kilometers beneath the summit. As the underground reservoir inflates, the pressure of the magma increases until the rocks above or on the sides of the reservoir crack and fail. Molten rock then pushes into these fractures, erupting at the surface (Plate 2) or halting underground as a shallow intrusion into the rift zones on the volcano's flanks.

There are two persistent zones of weakness in the sides of Kilauea: the East Rift Zone and the Southwest Rift Zone. As an injection of molten rock moves out into one of these rift zones, earthquake swarms outline the moving magma's progress. As the reservoir deflates, the summit subsides. Magma moving rapidly underground also causes volcanic tremor. By studying seismographs from many locations on the volcano and by measuring small changes in uplift, subsidence, expansion, or contraction

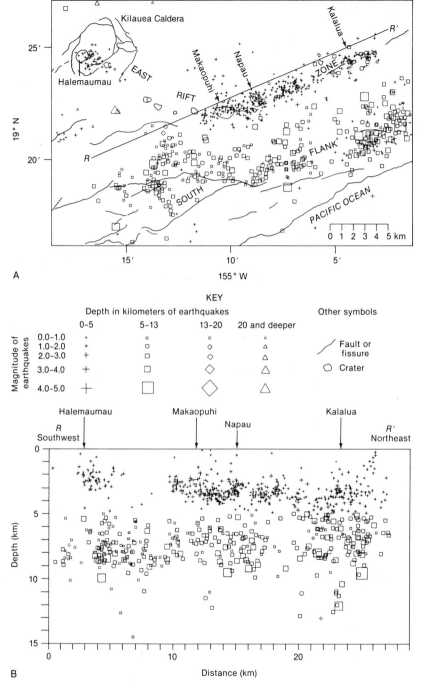

63 A. Locations of earthquakes beneath the summit and the East Rift Zone of Kilauea during January 1983. The shallow quakes between Makaopuhi and Kalalua were caused by the magma forcing its way into fractures at the start of the Pu'u O'o eruption. B. Earthquake locations projected into the cross section show that the magma conduit had its axis at a depth of about 3 kilometers. (From E. W. Wolfe et al., U.S. Geological Survey Professional Paper 1350, 1987, pp. 482–483.)

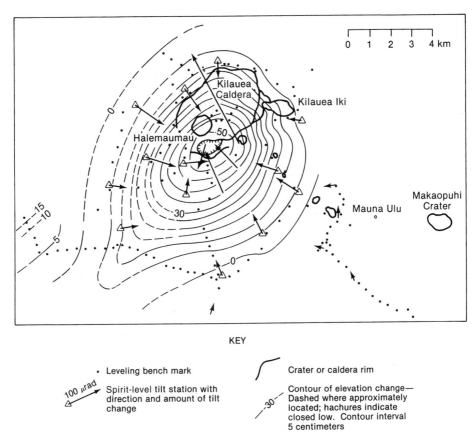

KEY

· Leveling bench mark   Crater or caldera rim

100 μrad / Spirit-level tilt station with   Contour of elevation change—
△   direction and amount of tilt   Dashed where approximately
         change   located; hachures indicate
                                                 closed low. Contour interval
                                                 5 centimeters

64 The summit of Kilauea subsided as magma drained into the east rift to supply the start of the Pu'u O'o eruption. The level surveys that determined this change were made between September 27 and October 5, 1982, and repeated between February 7 and 22, 1983. A continuously recording tiltmeter on the northwest side of Kilauea Caldera showed that nearly all this subsidence occurred between January 2 and 7, 1983. (From E. W. Wolfe et al., U.S. Geological Survey Professional Paper 1350, 1987, p. 488.)

of the volcano's surface, scientists routinely detect and map these underground movements of magma at Kilauea.

There were twelve intrusions of magma beneath the east rift from 1978 through 1982, only two of which erupted to the surface, one in 1979 and a very small event in 1980. Besides these rapid intrusions into new underground cracks, slower leakage from the summit reservoir into existing fractures caused additional swelling of the east rift in the area where the 1983 eruption fissures opened. This region was ready for a major eruption.

The fissures erupted intermittently between January 3 and 23; then lava emission stopped and the summit of Kilauea began to reinflate. Because short-lived volcanic events are common in Hawaii, many people thought the eruption was over, but on February 10, lava emission resumed, and from February 25 to March 4, vigorous lava fountains spewed from a vent about two-thirds of the way down the original fissure system. Again the eruption stopped, and tiltmeters at the summit, which had indicated subsidence, reversed direction and showed reinflation.

When summit inflation recovered to a critical level, would the eruption resume? The answer was yes. During its third episode, lava gushed forth in early April, and in a fourth episode, at a new site near the center of the original fissure system, lava fountains erupted in June. Between July 1983 and July 1986, this general eruption pattern—short periods of high lava fountaining from the same vent area lasting a few hours to days, separated by longer repose periods averaging about a month—was repeated an astounding 43 more times (Plates 1 and 5).

Although this three-year period of activity began to seem repetitious, geologists discovered interesting trends and changes in eruptive patterns that helped reveal the inner workings of Kilauea Volcano. Figure 65

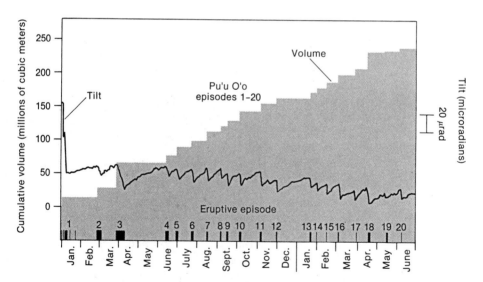

65 From 1983 until 1986, Pu'u O'o erupted intermittently. During its repose periods, the summit tilt showed a slow uplift, and during episodes of vigorous lava fountaining (shown in black with the episode number), the tilt indicated rapid subsidence of Kilauea's summit area. The substantial tilt loss at the beginning of the eruption (episode 1) was caused by the large volume of magma needed to fill the fracture system that initiated the eruption. (From E. W. Wolfe et al., U.S. Geological Survey Professional Paper 1350, 1987, pp. 490–491.)

shows both the cumulative volume of lava produced by the first 20 eruptive episodes and the subsidence–inflation cycles of the summit magma reservoir. Notice that the inward tilt of the summit during an eruptive episode is much steeper than the outward tilt during reinflation. This pattern indicates that the rate at which lava was erupted was much faster than the rate of supply of magma from depth.

The average rate of lava emission during the episodic phase of the eruption was 160 million cubic meters per year (Figure 66). As the magma erupted, most of the volcanic gases boiled out in vigorous fountains, although some gas stayed trapped in bubble holes, called *vesicles*, in the solidifying lava. To calculate the magma volume, the volume of the vesicles—about 25 percent—is subtracted. The average rate of magma production in the first years of the eruption was about 120 million cubic meters per year. Later in the eruption, the rate of lava emission slowed and then returned to its earlier levels. To date, the average rate of magma production has been close to 100 million cubic meters per year. Although the net tilt change during the past 14 years shows that the summit has subsided, the amount of magma withdrawn from beneath the summit has been small compared with the total amount of magma

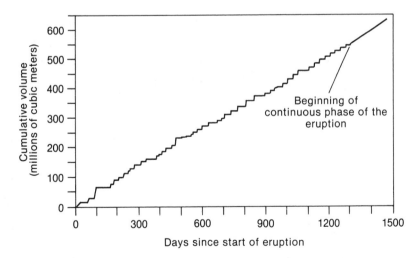

66 Cumulative lava production of the Pu'u O'o eruption, January 1983 to December 1986. In the episodic phase of the eruption, each fountaining event produced about 13 million cubic meters of lava. Each of the repose periods between episodes lasted about a month. After the continuous phase began in January 1986, the slow but steady emission of lava amounted to about 0.4–0.5 million cubic meters per day. Average lava production from the Pu'u O'o/Kupaianaha eruption from 1983 through 2003 was 125 million cubic meters per year. (Data from U.S. Geological Survey.)

erupted. Nonetheless, the supply of new magma from depth has largely kept up with the amount of lava erupted. This balance between the eruption rate and the resupply of magma implies that the eruption could be very long lasting.

As the early phase of the eruption progressed, the fountaining episodes became shorter, but the rate of lava emission increased (Figure 67), maintaining roughly the same amount of lava production per episode (about 13 million cubic meters). As the discharge rates increased, the lava fountains grew higher, exceeding 400 meters during episodes 24 to 30. This increase in vigor may have been caused in part by improved plumbing. That is, the magma coursing underground during each eruptive episode probably enlarged the conduits. Magma is hot enough (about 1150°C) to melt the wall rocks and enlarge the channels through which it moves. This is also thought to be the reason that a fissure eruption usually becomes localized at a single vent as a Hawaiian eruption progresses:

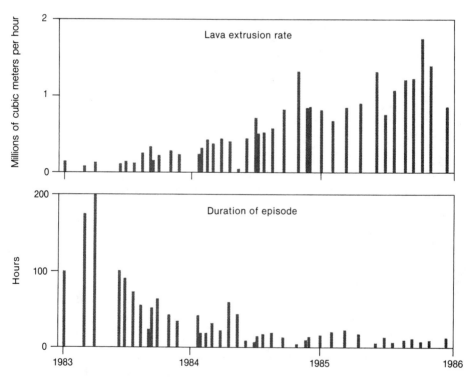

67   The rate at which lava was erupted and the height of the lava fountains generally increased as the Pu'u O'o eruption progressed, although the duration of each episode generally shortened. The net effect was to maintain a roughly equal amount of lava (about 13 million cubic meters) erupted in each episode. (From G. E. Ulrich, U.S. Geological Survey.)

where a fissure is narrow, less lava can flow through it, and the crack tends to heal shut as some lava cools and clings to the wall rock. Wider places in the fissure are enlarged as hot magma rapidly surges through them; hence, they increase in width and capture more and more of the flow.

Lava temperatures near the vents increased as the eruption progressed, from about 1120°C (episode 1) to 1140°C (episode 18), and the magnesium content of the basalt increased from about 6 percent to 7.5 percent. These changes are interpreted to mean that some of the magma erupting in the early episodes had been stored in the rift zone before eruption, cooling and precipitating olivine, a mineral rich in magnesium. Later in the eruption, most of the lava emitted was from hotter, more primitive magma stored and replenished in the summit reservoir.

With the vent localized in one spot along the fissure (episodes 4 through 47), the eruption began to build a major cinder and spatter cone around it (Figure 68). The cone was 20 meters high after episode 4, 130 meters high following episode 20, and by 1986 had reached a height of 255 meters, with a base about 1 kilometer in diameter. This imposing

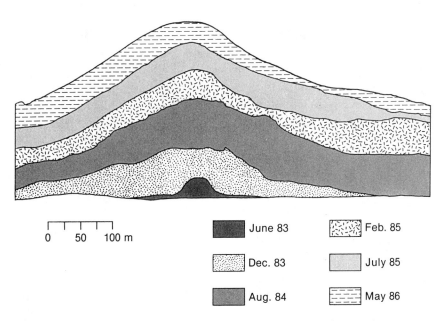

▓ June 83	▒ Feb. 85
░ Dec. 83	▢ July 85
▓ Aug. 84	▤ May 86

0   50   100 m

68 The growth of Pu'u O'o's cinder and spatter cone. Episodes 4 to 47 (1983–1986) were erupted from this vent and built the cone to a height of 255 meters with a base about 1 kilometer wide. These profiles are looking up the rift zone and are not vertically exaggerated. Since 1986, upward growth has stopped, lateral growth has increased, and collapse and slumping have greatly modified the cone's shape. (From Christina Heliker, U.S. Geological Survey.)

cone became a landmark visible from roads 30 kilometers away and was given the Hawaiian name Pu'u O'o, "Hill of the O'o Bird." Since 1986, the cone has been slumping into a lava lake inside its crater, thus losing elevation.

The lava falling back from the high fountains formed *a'a flows* (flows with rough surfaces of broken lava blocks) that moved away from the vent to the northeast, east, or southeast for 5 to 8 kilometers. Flows of this type are generally about 0.25 to 0.5 kilometer wide and 3 to 5 meters thick, fed by a central river of incandescent lava about 10 to 20 meters across. The central stream moves quite rapidly — 10 to 20 kilometers per hour — but as it spreads beneath the dark rubbly blocks near the front of the flow, it slows dramatically. The overall advance rates of the Pu'u O'o a'a flows were only a few hundred meters per hour. Although people had plenty of time to evacuate, several homes were crushed and burned as the relentless a'a flows covered parts of a rural subdivision 7 kilometers southeast of the vent.

The reason that the early eruptions at Pu'u O'o were episodic and geyserlike is not clear. Two factors seem to have been involved. The first was some sort of off–on valve in the conduit between the summit magma reservoir and the vent, which was sensitive to pressure changes. As the summit inflated and the magma pressure increased, new magma pushed through the conduit; then, as the summit deflated and the pressure dropped, the magma stopped flowing.

The second factor was the gas content of the magma. Lava appeared in the vent for several days before the fountaining started, and slow overflow of the vent preceded the high fountaining. Apparently a dense, gas-poor plug of lava had to overflow before the lighter, gas-rich lava that propelled the fountains could begin boiling. When the fountaining removed the top of the magma column, it reduced the pressure on the magma below, and it too flashed into boiling effervescence. This discharge continued at a high rate until the available magma and gas were exhausted.

The gases dissolved in magma are mainly water, carbon dioxide, and sulfur dioxide. At Kilauea they amount to less than 1 percent, by weight, of the magma. However, at high temperatures and low surface pressures, these gases vigorously boil out of the magma, and the expanded volume of the gas is many times the volume of the molten lava, driving the high lava fountains.

The high lava fountains produced large amounts of pumice and Pele's hair. *Pumice* is a quickly cooled volcanic rock puffed up by the expanding volcanic gases in tiny bubble holes; in a sense, it is solidified lava foam. The ratio of gas bubble holes to glassy volcanic rock is so high that some pumice lumps are as light as Styrofoam. Fine strands of volcanic glass that spin off from molten drops of lava whirling in the turbulent lava fountains are called *Pele's hair*. Both pumice lumps and Pele's hair were often blown several kilometers downwind from the lava fountains. Heavier lumps of solidified lava, called *cinders* or *scoria*, and still-molten lumps of spatter accumulated near the bases of the fountains and built the cinder and spatter cone around the vent. The Pu'u O'o cone grew highest on the southwest rim of the vent because the prevailing northeast trade winds piled most of the fallout from the lava fountains on the downwind side.

Pu'u O'o's seventeenth eruptive episode brought an answer to a long-standing scientific question. For years volcanologists had speculated on what the underground connections of Kilauea and nearby Mauna Loa volcanoes might be, and whether the activity of one would affect the other. Mauna Loa began a major eruption on March 25, 1984, and by March 30, a large flow of lava from a vent at 2900 meters elevation was headed toward the city of Hilo (Plate 6). Pu'u O'o was expected to erupt in late March, and everyone waited to see whether Mauna Loa's eruption would affect the activity at Kilauea, or vice versa. Episode 17 arrived on schedule, with 100-meter-high fountains that lasted for 23 hours and poured out 10 million cubic meters of lava. It was rare and marvelous to see both volcanoes erupting at the same time.

Even though the vent at Pu'u O'o was more than 2000 meters lower than the vent at Mauna Loa, neither eruption had any apparent effect on the other. Although both Kilauea and Mauna Loa volcanoes may get their magma from the same general deep source, it is clear that there is no direct hydraulic connection between the shallow magma reservoirs of the two volcanoes.

Three and a half years after the start of the Pu'u O'o eruption, episode 48 in July 1986 brought a significant change. Fractures opened both up-rift and downrift from Pu'u O'o; a new vent became established about 3 kilometers downrift, and activity was now focused there. Smooth-surfaced lava flows called *pahoehoe* issued slowly and steadily from this new vent, building a broad, low pile of solidified lava around it and ponding the upwelling magma into a 100-meter-wide active lava lake. Lava sometimes overflowed the rim of the lake, but more often it escaped through a tunnel to feed pahoehoe flows that slowly advanced through lava tubes beneath their surfaces.

In contrast to the earlier episodic pulses, the downrift eruption became nearly continuous, pouring out lava at a slowly decreasing rate for more than six years. The vent and the lava shield built up around it during this second phase of the Pu'u O'o eruption was named Kupaianaha, the Hawaiian word for "mysterious." Although the pahoehoe lava flows from this continuous phase of the eruption advanced more slowly than did the a'a flows from the earlier intermittent eruptions, they slowly but inexorably moved down the southeast flank of the rift to the sea. The flows reached the ocean in November 1986, destroying rural homes and covering a highway in their path. Many more homes and a National Park Visitor Center were destroyed in later flows from Kupaianaha.

After a period of decreasing output of lava, this second phase of the eruption came to an end in early 1992. Saddened by the destruction, most people in Hawaii hoped the eruption was over for good, but that was not to be the case. About a week later, a new vent opened on the uprift flank of the Pu'u O'o cone. Nearly continuous flows from this third phase of the eruption again moved slowly toward the sea, inundating new areas on the west side of the still-growing flow field (Figure 69). A lava lake formed deep inside the old crater of Pu'u O'o, and its surface level now fluctuates with the emission rate of lava from the new vent area.

The molten rock moves in lava tubes beneath the surface of pahoehoe flows from the vent area to the sea, a distance of about 7 kilometers. Sometimes the tubes become blocked with chunks of lava crust. When this happens, new flows break out from holes in the lava tubes, called *skylights*, uphill from the blockage. Often the tubes clear themselves and the flows to the sea are reestablished. Lava-generated steam sometimes causes small explosions as the red-hot rock slips into the sea, but usually a pillar of steam is all that marks a remarkably docile meeting of fire and water (see Figure 17).

As the molten lava is suddenly quenched by the sea, some of it shatters into sand-sized fragments. This new sand is swept along by shoreline currents and forms black sand beaches where once the surf pounded on bare rock ledges. These new beaches accumulate in embayments along the coast, well beyond the edges of the flows entering the sea.

The billowing surface of the pahoehoe flows that form most of the land that has been added to the Big Island looks like easy walking, but its appearance is deceptive. Not only is it possible to break through the surface into a lava tube, but large sections of this newly added land can suddenly slump into the sea. This happens without warning and poses a major danger to volcano watchers trying to get a better view of the molten lava meeting the ocean.

Why did the Pu'u O'o eruption become slow and steady after the first phase? Part of the explanation may be that most of the volcanic gas was and still is venting from the Pu'u O'o cone, and not enough gas remains in the magma at the erupting vent to drive the high lava fountains.

The gases also seem to be responsible for another phenomenon, a haze of volcanic fume locally called *vog* (volcanic fog) that has plagued

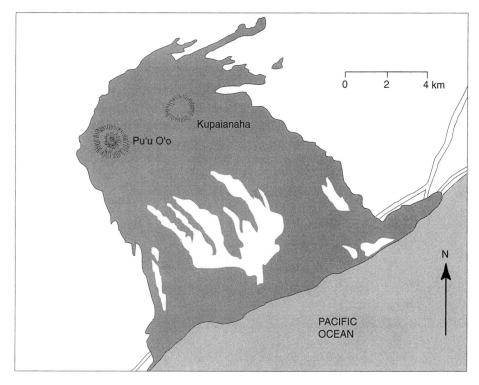

69 Lava flows from the long-lasting Pu'u O'o/Kupaianaha eruption on the east rift of Kilauea Volcano. From January 1983 to January 2003, 111 square kilometers of land (8 percent of Kilauea) was covered by flows, mainly from the Pu'u O'o and Kupaianaha vents. The total volume of lava was nearly 2.5 cubic kilometers, and the flows destroyed 189 buildings, including a National Park Visitor Center. Fourteen kilometers of road along the coast were covered, and 2.2 square kilometers of new land was added to the island of Hawaii. (Drawing by Richard Hazlett.)

parts of the Big Island since the lava began entering the ocean in late 1986. The ingredients of this vog come not only from the sulfur gases from Pu'u O'o and the active vent areas, but also from an aerosol of hydrochloric acid droplets formed in the steam plume where the lava meets the sea. The trade winds generally blow the vog out to sea near the eruption, but it usually eddies back on the west side of the island—much to the irritation of residents used to warm, clear breezes and unobscured views of the sunset.

The latest event of the eruption is occurring as we write this chapter in early June 2004. After nearly a year's hiatus during which the ongoing lava flows from Pu'u O'o stopped short of the ocean, they have again started pouring into the sea (Plate 8). The spectacular photos in the local papers and on the Hawaiian Volcano Observatory Web site are luring thousands of visitors to hike over the cooled but barren earlier flows to see this fantastic sight in person.

Although the Pu'u O'o eruption is the longest-lasting and largest rift eruption in the recorded history of Kilauea, it is not unprecedented. Summit eruptions of Kilauea were nearly continuous from 1823 to 1924, and one prehistoric east rift eruption appears to have been much more voluminous—and probably longer lasting—than the present one. Each new eruption leads to important insights into how volcanoes work. But as quickly as old questions are answered, new ones arise. Why does the rate of lava emission vary? How do the lava tubes form, block up, and reopen? Why does some of the new land added to Hawaii in this eruption slump into the sea in sections larger than a football field? How does the hydrochloric acid aerosol form in the steam plumes where the lava enters the sea? There is much for the current group of scientists at the Hawaiian Volcano Observatory to learn. There always will be.

Links

Center for the Study of Active Volcanoes
http://www.uhh.hawaii.edu/~csav/

Hawaii Center for Volcanology
http://www.soest.hawaii.edu/GG/hcv.html

Hawaii Scientific Drilling Project
http://icdp.gfz-potsdam.de/sites/hawaii/news/news.html

Hawaii Volcano Geochemical Data Sets
http://www.bishopmuseum.org/research/natsci/geology/geochem.html

Hawaii Volcanoes National Park
http://www.nps.gov/havo/home.htm

Hawaii Volcanoes NP Field Trips
http://www.botany.hawaii.edu/faculty/bridges/bigisland/field_sites.htm

Hawaiian Volcano Observatory
http://hvo.wr.usgs.gov/

How Volcanoes Work
http://www.geology.sdsu.edu/how_volcanoes_work/

U.S. Geological Survey Pu'u O'o Fact Sheet
http://geopubs.wr.usgs.gov/fact-sheet/fs144-02/

Volcanology, Geochemistry & Petrology, University of Hawaii
http://www.soest.hawaii.edu/GG/vgp.html

Sources

Decker, Robert, Thomas Wright, and Peter Stauffer, eds. *Volcanism in Hawaii*. 2 vols. U.S. Geological Survey Professional Paper 1350, 1987.

Hazlett, Richard W. *Geological Field Guide: Kilauea Volcano*. Hawaii National Park: Hawaii Natural History Association, 2002.

Hazlett, Richard W., and D. W. Hyndman. *Roadside Geology of Hawaii*. Missoula, MT: Mountain Press, 1996.

Heliker, Christina, D. A. Swanson, and T. J. Takahashi, eds. *The Pu'u O'o-Kupaianaha Eruption of Kilauea Volcano*. In *Hawaii: The First 20 Years*. Washington, DC: U.S. Geological Survey Professional Paper 1676, 2003.

Macdonald, Gordon A., Agatin T. Abbott, and Frank L. Peterson. *Volcanoes in the Sea*. 2nd ed. Honolulu: University of Hawaii Press, 1983.

Mattox, Tari N., Jeff Sutton, and Tamar Elias. "Where Lava Meets the Sea, and Volcanic Gases Create Air Pollution." *Earthquakes and Volcanoes* 24 (1993).

Tilling, R. I., and J. J. Dvorak. "Anatomy of a Basaltic Volcano." *Nature* 363 (1993): 125–133.

Tilling, Robert, C. Heliker, and T. L. Wright. *Eruptions of Hawaiian Volcanoes: Past, Present, and Future*. U.S. Geological Survey, 1987.

CHAPTER 8

Hot Spots

70   Lava lake in Hawaii. (Photograph by D. H. Richter, U.S. Geological Survey.)

The close connection between the edges of the moving plates of the Earth's crust and volcanoes has been emphasized in earlier chapters. Why, then, is the island of Hawaii, with Kilauea and Mauna Loa, two of the world's most active volcanoes, right in the middle of the Pacific Plate (Figure 71)? Only 5 percent of the world's active volcanoes are located within plates, but even so, there must be some reason for their existence.

There are some obvious differences between the Hawaiian volcanoes, which form a linear belt or chain, and volcanoes located on plate margins. Topography is one. A profile section across the Hawaiian chain is completely different from a cross section of a spreading ridge or a subduction zone (Figure 72).

Another major difference is the age distribution of Hawaiian volcanoes compared with that of plate-margin volcanoes. For more than 150 years, geologists have recognized that the islands in the Hawaiian chain are older to the northwest. Since 1800, eruptions have occurred only on Hawaii, at the southeastern end of the chain; the islands to the northwest are lower and more eroded. This island chain extends to Midway Island, 2500 kilometers northwest of the Big Island of Hawaii, where only a coral atoll cap covering a submerged volcanic peak marks the largely submarine ridge. The ages of the volcanic rocks of the Hawaiian Islands, obtained by radiometric dating, confirm this picture. All the rocks on the Big Island are less than 1 million years old; most of the rocks on Oahu, the location of Honolulu and Pearl Harbor, are 2 million to 3 million years old; and the ancient lava flows on Kauai, northwest of Oahu, are about 5 million years old (Figure 73).

This age distribution pattern is strikingly different from that of plate-margin volcanoes. In subduction zones and on mid-ocean ridges, young volcanoes form in lines near and along the margins of the plates, and on the mid-ocean ridges there is a progressive aging of the volcanic rocks on either side of the active crest. The Hawaiian chain is just the opposite: the volcanoes are older along the crest of the chain instead of across the crest of the ridge.

Earthquakes—close companions of active volcanoes—also show an unusual pattern in Hawaii. Although earthquakes occur along the entire lengths of subduction zones and oceanic ridges, in Hawaii they are com-

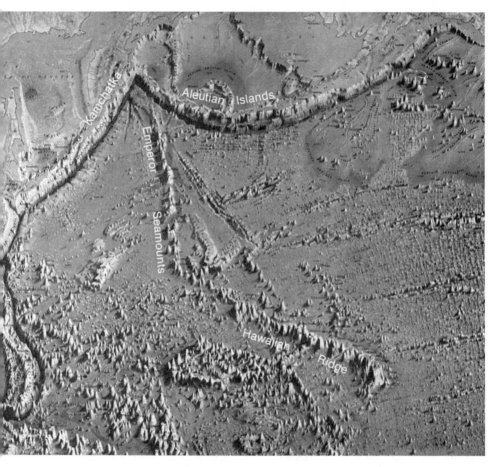

71 The Hawaiian Islands rise from the southeastern end of a 6000-kilometer-long dogleg chain of seamounts. The portion north of the bend is called the Emperor Seamounts; that south of the bend is called the Hawaiian Ridge. (Map from *World Ocean Floor Panorama* by Bruce C. Heezen and Marie Tharp, initiated and supported by the Office of Naval Research. © Marie Tharp. Reproduced by permission of Marie Tharp. All rights reserved.)

mon only at the southeastern end of the chain, mostly beneath the Big Island.

The data from more than 30,000 microearthquakes occurring near Kilauea Volcano on Hawaii, as well as the pattern of tilting that indicates the swelling and contraction of a shallow magma chamber, suggest an interesting subsurface model of Hawaiian volcanoes (Figure 74). Charting the locations of earthquakes at depth indicates the conduits through which molten rock forces its way upward. The propelling force is the buoyancy of the lighter molten rock surrounded by denser crystalline rocks. As the magma rises—probably in many separate cracks rather than

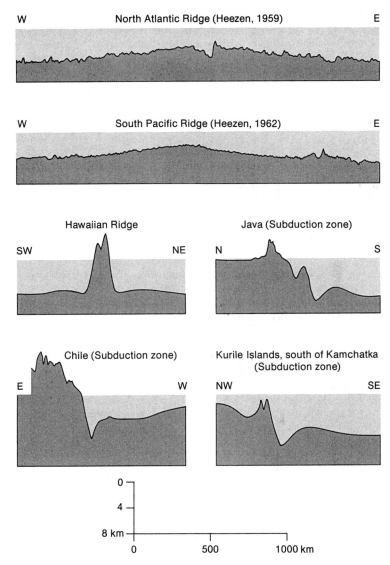

72 Comparison of topographic profiles of subduction zones, mid-ocean ridges, and Hawaii. The vertical scale is exaggerated forty times. The island arcs and compressional mountains of subduction zones are basically asymmetrical, whereas the mid-ocean ridges and Hawaii are symmetrical around a central axis. The width of the mid-ocean ridges is a result of seafloor spreading.

a continuous conduit—the hard surrounding rocks are broken open by changing pressures and temperatures, causing earthquakes.

The deep source of magma appears to be within the plastic layer 50 to 60 kilometers beneath the Pacific Plate, the maximum depth of the earthquakes. Diagrams of the locations of the intermediate to deep earth-

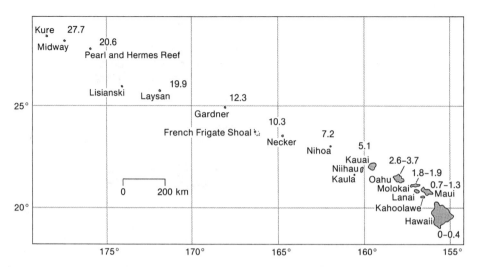

73 The ages of islands in the Hawaiian chain in millions of years. The farther northwest one travels, the older the islands are. The apparent rate of movement away from the Hawaiian hot spot (deep source of magma) is about 9 to 10 centimeters per year. (Data from D. A. Clague and G. B. Dalrymple, U.S. Geological Survey.)

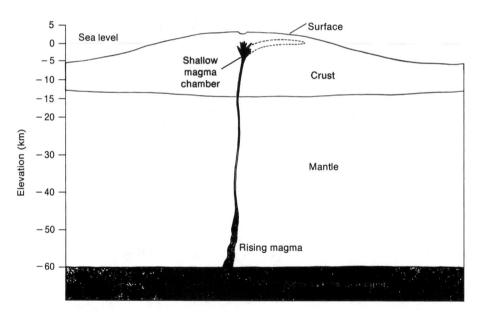

74 A schematic cross section of an active Hawaiian volcano. The vertical scale is exaggerated two times. The dotted line extending to the right from the shallow magma chamber is an inferred zone of feeder dikes connecting it to a rift zone. The deep source of magma, below 50–60 kilometers, and the shallow storage chamber at 3 to 6 kilometers beneath the surface are the vital parts of the active system. (After Jerry Eaton, U.S. Geological Survey.)

quakes, 10 to 60 kilometers beneath Kilauea, have outlined over a period of years a zone shaped like an inverted funnel 5 to 50 kilometers in diameter and 50 kilometers high, which allows the deep magma to rise into a shallow magma chamber 3 to 6 kilometers beneath the summit. Even though this shallow chamber is bounded by a region of many microearthquakes, there are almost no earthquakes inside the chamber. Most earthquakes are thought to originate from the cracking and slipping of brittle rocks that suddenly fracture. Plastic or liquid rock in a magma chamber can deform without breaking and thus produces no earthquakes (Figure 75).

The reason that magma chambers often form at these shallow depths is explained by the concept of *neutral buoyancy*, an idea championed by Mike Ryan of the U.S. Geological Survey. Below 6 kilometers the surrounding rocks are more dense than magma, thus causing it to rise; above 3 kilometers the surrounding rocks are less dense than magma, impeding its rise.

The tilt pattern near the summit of Kilauea supports this magma chamber model. Just as the pattern of tilting on the surface of an inflating balloon is related to the radius of the balloon, the pattern of tilting at Kilauea points to a pressure center about 4 kilometers deep. Continuing this balloon analogy, the deep source of magma is the pump; the inverted-funnel conduits are the connecting hoses, and the shallow magma chamber is the balloon.

The pattern of Kilauea's slow outward tilting (inflation) before an eruption and rapid inward tilting (deflation) during a flank eruption indicates a slow, almost continuous movement of magma from the deep source to the shallow chamber and rapid and intermittent eruptions from the shallow chamber to the surface. Because the rate of movement of magma from depth to the shallow chamber often does not keep pace with the surface eruption rate, many eruptions stop until the shallow chamber has been recharged. The eruption at Pu'u O'o has been an exception. Since the steady-state lava emission began there in 1986, most of the lava has been slowly but continuously resupplied from depth. There has been a net deflation of Kilauea's summit during the 21 years of the current eruption, but the calculated volume of magma that would cause this deflation is small compared with the total volume of lava that has erupted.

How do these processes relate to the origin and continuing vigor of the Hawaiian volcanic chain? There must be something unique about the location of the island of Hawaii. Somehow, an ongoing magma source must be available at this location. Tuzo Wilson, a Canadian geophysicist, suggested an elegantly simple explanation that fits neatly with the

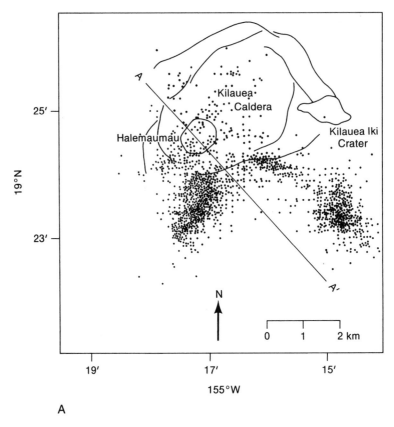

75 A. A map view of the locations of shallow microearthquakes underneath Kilauea Volcano in Hawaii. Earthquakes are caused by rock fracture, so zones of no earthquakes are regions of low stress or very low strength. Since high stresses are common near most active volcanoes, the lack of earthquakes in a region surrounded by them suggests a plastic zone occupied by magma. The dense clusters of microearthquakes shown on the map indicate the beginning of the rift zones.

movement of the Pacific Plate. He envisioned an upwelling of hot, plastic rock originating deep beneath the plate. This upwelling, or *plume*, rises by very slow convection and causes some melting because of decompression as the hot rock rises in the mantle. The plume and its molten top, or *hot spot*, stay in the same location for millions of years as the plate slides over this continuing source of magma. A volcano grows over the hot spot and drifts away with the slowly moving plate, to be replaced by a new volcano growing over the same hot spot, like smoke signals drifting in a gentle wind (Figure 76). Most geologists now accept the hot-spot idea.

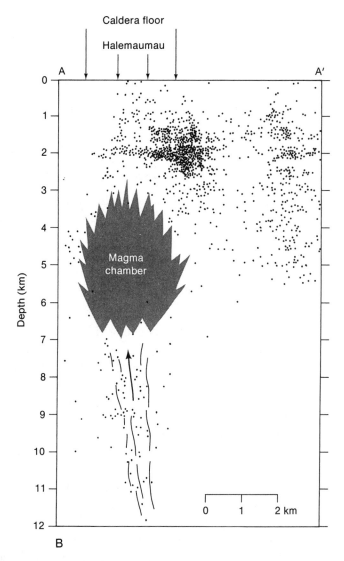

75 B. Cross section of Kilauea Caldera to a depth of 12 kilometers. (Data from Robert Koyanagi, U.S. Geological Survey.)

The problem, of course, is explaining the character of such a thermal plume. What keeps it going for millions of years, generating new magma at a relatively fixed spot in the Earth? One prominent idea is that the heat source of the hot spot is in the deep mantle, perhaps even at the boundary between the core and the mantle. This implies that the magma supplying hot-spot volcanoes has an even deeper source than does that supplying the volcanic belts along the plate margins (Figure 77).

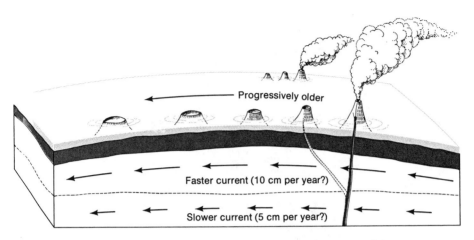

76 The original diagram used by Tuzo Wilson to suggest the origin of the Hawaiian Islands. The plate moves slowly over a relatively fixed source of magma, forming a new volcano as the older volcanoes are carried downstream. (From J. Tuzo Wilson, "Continental Drift." Copyright © 1963 by Scientific American, Inc. All rights reserved.)

This deep-plume hypothesis has been challenged by a growing group of geologists and geophysicists in the past few years, mainly on the basis that seismic tomography does not reveal narrow plumes of hotter-than-average material in the deep mantle. (*Seismic tomography* measures the variation in seismic wave velocities within the mantle in a way analogous to a sonogram of an unborn baby in the womb or an X-ray or magnetic resonance scan of the interior structures of the body.) These shallow-plume or anti-plume advocates favor shallower sources of magma that are maintained by ongoing surface-extension processes or magma formed by localized regions of water-rich mantle that have resulted from old subducted slabs. The debate is lively, much as it was in the early days of plate tectonics, and it can be followed as it progresses through the Web site links listed at the close of this chapter.

The Hawaiian hot spot, whether of deep or shallow origin, seems to be wide enough to feed three volcanoes at the same time—Mauna Loa, Kilauea, and the submarine volcano Loihi, off the south coast of the Big Island. It also appears that the Hawaiian hot spot must have been generating magma for at least 80 million years in order to form the Hawaiian Ridge and Emperor Seamounts (see Figure 71).

If the hot-spot theory is correct, then the aging of volcanic rocks away from the hot spot should match the rate of plate movement. The magnetic stripe pattern on the ocean floor indicates that the Pacific Plate is moving northwest at about 10 centimeters per year. If Kauai has drifted northwest during the 5 million to 6 million years since its lavas erupted,

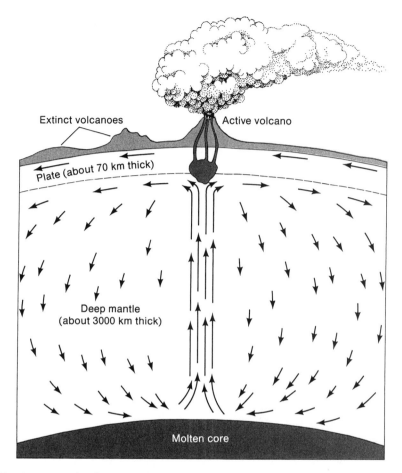

77 Deep mantle plumes generated by slow convection currents are one explanation for the relatively fixed position of hot-spot volcanic sources. The concept illustrated here was originated by Jason Morgan. (From G. B. Dalrymple, E. A. Silver, and E. D. Jackson, " Origin of the Hawaiian Islands," *American Scientist*, March 1973, p. 306.)

it should be 500 to 600 kilometers away from where it first formed, at the present location of the Big Island. In fact, the actual distance is 560 kilometers, a striking verification of Wilson's hot-spot idea.

Beyond Midway Island, the Hawaiian chain is a submerged line of seamounts continuing northwest for another thousand kilometers. After that, it meets a submerged mountain chain, the Emperor Seamounts, which trends north–northwest another 2500 kilometers toward Kamchatka. The Emperor Seamounts are a dogleg of the Hawaiian chain, formed when the drift of the Pacific Plate was more northward than its present northwestward movement. The change in direction of plate movement, as estimated from the location of the bend, occurred about 43 million years ago. Of course, an alternative explanation is that the hot-spot source, rather than the plate, changed direction.

The hot-spot explanation of the Hawaiian chain suggests that a future island will form to the southeast of the Big Island. The Loihi seamount, 28 kilometers southeast of the island of Hawaii, looks as if it will be that next island. Loihi is 970 meters below sea level, with two deep craters indenting its summit. Divers in submersibles have discovered pillow-lava flows on Loihi that appear to be very young. Warm springs with dissolved carbon dioxide have also been observed and sampled there.

Swarms of small earthquakes underneath Loihi were recorded in 1971–1972 and in 1975, and they may have been related to submarine eruptions, though this correlation is by no means certain. During the summer of 1996, the most energetic earthquake swarm ever recorded from Loihi took place. Over a period of 20 days, the Hawaiian Volcano Observatory recorded more than 4000 Loihi earthquakes; 95 had magnitudes greater than 4, and the largest quake was a magnitude 5. Submariners at the University of Hawaii had been waiting for this kind of activity at Loihi, and dives were made in the submersible *Pisces* as the earthquake swarm waned. A new collapse crater was observed near the summit of Loihi, indicating that magma may have drained away from beneath that area, but no red lava or active eruptive vents were seen. The divers also encountered waters filled with fine white "dust" that reduced visibility to less than 2 meters. Whether Loihi is vigorously active or not, it will probably be tens of thousands of years before it becomes the next Hawaiian island.

The great strength of the plate tectonics and hot-spot concepts is that their predictions are substantiated by clues found in so many places around the world. The plates have left their tracks in the form of volcanic scars and magnetic fields, and those tracks reveal a complex yet consistent global pattern.

The number of hot spots worldwide is still a matter of controversy (Figure 78). At least two seamount chains in the South Pacific indicate other major hot spots beneath the Pacific Plate. Yellowstone National Park, on the North America Plate, has been suggested as an active continental hot spot, with the Snake River volcanic plain its earlier track. The Galápagos Islands on the equator off the west coast of South America is another group of very active volcanoes of probable hot-spot origin.

The hot-spot idea even offers a solution to the puzzle of why Iceland is the only major part of the Mid-Atlantic Ridge out of water. Iceland is

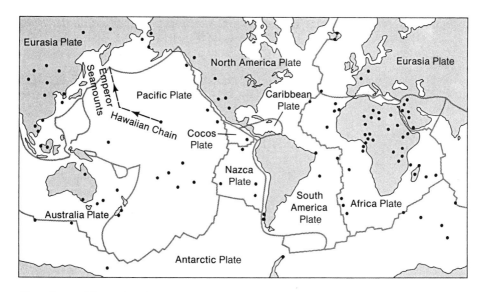

78   The world's current hot spots and their relation to the major plates. Hot-spot tracks, as illustrated by the Hawaii–Emperor trend, reveal past plate motions. Some geologists prefer a smaller number of hot spots. (After Kevin C. Burke and J. Tuzo Wilson, "Hot Spots on the Earth's Surface." Copyright © 1976 by Scientific American, Inc. All rights reserved.)

apparently a hot spot that straddles the ridge and has volcanoes of dual origin—both hot-spot and rift volcanoes. This is an attractive explanation of the extra volcanism that has built Iceland above the sea.

No islands of similar origin could be more unlike than Hawaii and Iceland. Iceland is a rugged, nearly treeless land of harsh climate and stark beauty, inhabited by strong people of Viking heritage who have stalwartly maintained their culture and independence. Hawaii is a tropical garden, soft and seductive, inhabited by tourists who have almost submerged its Polynesian heritage. But both islands owe their distinctive beauty, power, and mystery to their origins in volcanic fire.

Links

Flood Basalts
  http://www.geolsoc.org.uk/template.cfm?name=fbasalts

The Formation of the Hawaiian Islands
  http://www.soest.hawaii.edu/GG/HCV/haw_formation.html

How Volcanoes Work
  http://www.geology.sdsu.edu/how_volcanoes_work/

Large Igneous Provinces
  http://www.largeigneousprovinces.org/

Loihi Volcano
http://www.soest.hawaii.edu/GG/HCV/loihi.html

The Mantle
http://132.156.108.210/personalpages/roddick/mantle_e.htm

Origin of "Hotspot" Volcanism
http://www.mantleplumes.org/

Yellowstone Hotspot
http://www.dur.ac.uk/g.r.foulger/Offprints/Yellowstone.pdf

Yellowstone Volcanic System
http://volcanoes.usgs.gov/yvo/products.html

Wot, no plumes?
http://www.geolsoc.org.uk/template.cfm?name=NakedEmperor

Sources

Anderson, D. L. "Top-Down Tectonics?" *Science* 293 (2001): 2016–2018.

Anderson, D. L., T. Tanimoto, and Y. Zhang. "Plate Tectonics and Hotspots: The Third Dimension." *Science* 256 (1992): 1645–1651.

Burke, Kevin, and J. Tuzo Wilson. "Hot Spots on the Earth's Surface." *Scientific American*, August 1976.

Christiansen, Robert L., G. P. Foulger, and John R. Evans. "Upper-Mantle Origin of the Yellowstone Hotspot." *Geological Society of America Bulletin*, 114 (2002): 1245–1256.

Clague, David, and C. Brent Dalrymple. "The Hawaiian–Emperor Volcanic Chain." In *Volcanism in Hawaii*. U.S. Geological Survey Professional Paper 1350, 1987.

Courtillot, V., A. Davaille, J. Besse, and J. Stock. "Three Distinct Types of Hotspots in the Earth's Mantle." *Earth and Planetary Science Letters* 205 (2003): 295–308.

Hamilton, Warren B. "An Alternative Earth." *GSA Today* 13, no. 11 (2003): 4–12.

Vink, Gregory, Jason Morgan, and Peter Vogt. "The Earth's Hot Spots." *Scientific American*, April 1985.

CHAPTER 9

Lava, Ash, and Bombs

79 Pahoehoe lava flow, Hawaii. (Photograph by U.S. Geological Survey.)

Volcanoes are dark windows to the Earth's interior. Because their products are the only direct samples of the composition of the Earth's deeper levels, the aspect of volcanology that studies these materials has received considerable attention. Besides helping us determine what the Earth is made of, the character of successive layers of volcanic deposits can reveal the type and sequence of prehistoric eruptions and provide a broader picture of volcanic events than present activity can.

Although most people think of lava flows as the main products poured forth from volcanoes, volcanic ash and larger solid fragments—called volcanic cinders, bombs, and blocks—are actually the major products of observed eruptions (Table 1). Lumping together all sizes of solid fragments,

TABLE 1   Forms of volcanic products

Form	Name	Characteristics (dimensions)
Gas	Fume	
Liquid	*Lavas* A'a Rough, blocky surface Pahoehoe	Smooth to ropy surface
Solid	*Airfall fragments* Dust Ash Cinders Bombs Blocks	$<\frac{1}{8}$ mm $\frac{1}{8}$–2 mm 2–64 mm >64 mm plastic >64 mm solid
	Pyroclastic flows	Flows fluidized by hot gases
	Mudflows	Flows fluidized by rainfall, melting ice and snow, or ejected crater lakes
	Avalanches	Thick, hummocky deposits of broken rock

geologists use the term *pyroclastic deposits* for this volcanic debris (*pyroclastic* literally means "fire fragment"). Pyroclastic materials are derived from three sources: magma that is cooled and broken into fragments by expanding gases at the time of eruption; fragments of old crater walls or magma conduits that are ripped loose in explosive eruptions; and clots of liquid lava thrown into the air that cool and solidify during their flight.

Pyroclastic materials are classified by the general size of the fragments. *Volcanic ash* can be as fine as flour but usually is more gritty, with particles up to the size of rice; *volcanic cinders* (also named *lapilli—* "little stones" in Italian) are larger than particles of ash and can be as big as golf balls; and *volcanic blocks* can be anything larger than cinders, including chunks the size of a house. *Volcanic bomb* is a special term for a cinder- or block-sized clot of liquid lava thrown from an erupting vent (Figure 80). The brilliant arcs on time-lapse photographs of volcanic eruptions are the traces of volcanic bombs in flight.

Pyroclastic materials are deposited in two different ways: by pyroclastic fall or by pyroclastic flow. Explosive volcanic eruptions often lift and

80 A large volcanic bomb from Mauna Kea Volcano, Hawaii. This 50-kilogram specimen shows the twisted flow lines often acquired while these globs of hot plastic lava are spinning in flight. The pencil at the bottom shows the scale.

hurl fragments to great heights. As the particles fall back to the Earth, they form a distinct layer or layers that blanket the slopes of the land. An Indian legend about a pre-1980 eruption of Mount St. Helens refers to a pyroclastic fall of volcanic ash as "gray snow that didn't melt." The falling process allows for some sorting of the debris: the coarser fragments fall first and nearby, while the fine ash is winnowed away to fall last, sometimes at great distances. The deposit from a pyroclastic fall can be called just that, but for brevity some geologists use the term *airfall deposit*. An airfall deposit can usually be identified by the way it blankets the topography and by the way its particles are sorted (Figure 81).

81 Volcanic ash layers in a road cut on Oshima Volcano in Japan. Airfall deposits are characterized by distinct layers that follow the slope of the ground surface on which they fell. Two generations of ashfalls are seen here, separated by an erosional surface. The beds are not folded; the inclined layers are in the original attitude in which they accumulated.

Sometimes an explosive eruption produces a cloud of volcanic debris so full of fragments that it is too heavy to rise very high into the atmosphere. Such an emulsion of gas and fragments forms a *pyroclastic flow*, the most dangerous kind of volcanic hazard. Pyroclastic flows can travel at speeds of more than 100 kilometers per hour, flattening and burning almost everything in their path. Small pyroclastic flows often race down the valleys on a volcano's flanks, but larger masses that are expelled at high speeds or those accelerated by falling can sweep over small hills or across large flat areas. Pyroclastic flow deposits often pond in low-lying areas after they lose speed. Large, thick pyroclastic flow deposits are distinct from airfall deposits because they exhibit only vague layering and almost no sorting of the finer and coarser fragments. Road cuts in airfall deposits show distinct bands of coarse and fine layers, often of different colors, while cuts in thick pyroclastic flows show massive deposits that look like pink or buff concrete.

The nomenclature of *pyroclastic flows* (the process) and *pyroclastic deposits* (the products) is both complex and confusing, because the descriptive terms come from several languages. In addition, these flows are difficult to observe, as they are often deadly to close witnesses but hidden from more distant witnesses by the clouds of volcanic ash that shroud them. The challenge is to understand the nature of pyroclastic flows from their deposits, which often leads to conflicting opinions. Here we condense and paraphrase the descriptions of various types of pyroclastic flows and deposits by Peter Francis in his excellent book *Volcanoes: A Planetary Perspective*.

Pumice flows deposit *ignimbrites* (Latin for "fire cloud rocks"), which are rocks or unconsolidated deposits of various-sized fragments of pumice. *Nuées ardentes* (French for "glowing cloud") deposit blocks and ash composed of fragments more dense than pumice. *Pyroclastic surges* are flows of gas and fragments less dense than either pumice flows or *nuées ardentes*, and they leave thin but extensive deposits with cross-bedded layering (Figure 82). The general terms for all these flows and deposits are *pyroclastic flows* and *pyroclastic deposits*. Examples help distinguish these terms, and Francis provides many.

Huge pumice flows can form by the collapse or fallback of an erupting cloud column so laden with fragments that the convective lift from its internal heat can no longer sustain its rise. The 7700-year-old ignimbrite that spewed from Mount Mazama as the caldera of Oregon's Crater Lake was forming is an example of a massive column-collapse pyroclastic flow. The volume of the Mazama ignimbrite is tens of cubic kilometers, and it extends as far as 50 kilometers around Crater Lake. Some thick ignimbrites that are very hot when deposited consolidate into

82 Pyroclastic surge deposits from the 1790 explosive eruption of Kilauea
Volcano in Hawaii. The cross-bedding shows that wind velocities in the surges
rose to speeds high enough to erode layers of previously deposited pyroclastic
fragments. Transport direction was from left to right. The hammerhead (upper
left) indicates the scale. (Photo by Robert L. Christiansen, U.S. Geological Survey.)

welded *tuffs*—hard, dense rocks containing pumice lumps that collapse into small, flat obsidian disks.

Base surges are expanding rings of turbulent mixtures of fragments and gas that surge outward at the base of explosion columns. The base surge that formed at the nuclear test on the Pacific island of Bikini in 1946 rolled outward beneath the mushroom cloud for a distance of 4 kilometers at an initial speed of 50 meters per second. That was the speed of the ring's expansion; the turbulent internal wind speeds were probably much higher. The fluid dynamics of a base surge appear to be similar to the formation and radial growth of a smoke ring.

Volcanic eruptions involving the explosive interaction of magma and water seem to be the ones that generate base surges, which are now called pyroclastic surges when related to volcanic explosions. Surge deposits are generally small in volume, extensive in area, cross-bedded like sand dunes, and do not extend more than a few kilometers from their vents. The 1965 eruption of Taal Volcano in the Philippines, studied by Jim Moore of the U.S. Geological Survey, provides a good example of pyroclastic surges and their deposits. The eruption took place near the shore of Volcano Island, a 5-kilometer-wide volcanic island in Lake Taal. During a two-day period, many horizontal blasts swept across the lake from the vent area. Trees near the vent were uprooted; farther away, the bark and wood on the sides of trees facing the vent were eroded away, but on their lee sides they were unaffected. Damage extended as far as 6 kilometers from the vent. More than 4 meters of dune-shaped surge deposits accumulated within a kilometer of the vent, but their thickness diminished rapidly to less than 1 centimeter 8 kilometers away. Mud was plastered on tree trunks and other vertical surfaces in the more distant parts of the devastated area, indicating that the surge at those locations was wet and below boiling temperature (Figure 83).

The name *nuée ardente* was first given by the French geologist Alfred Lacroix to the pyroclastic flow from Mont Pelée that annihilated the town of St. Pierre, Martinique. The flow was caused by the explosive breakup of a growing lava dome at the summit of the volcano, which then swept down on the city (Figure 84). One of the few witnesses to survive, Purser Thompson on the steamship *Roraima*, which was anchored in the harbor, described the eruption:

> For hours before we entered the roadstead, we could see flames and smoke rising from Mont Pelée. . . . There was a tremendous explosion about 7:45 o'clock, soon after we got in. The mountain was blown to pieces. . . . The side of the volcano was ripped out, and there hurled straight toward us a solid wall of flame. It sounded like thousands of cannon. The wave of fire was on us and over us like a lightning flash. It was like a hurricane of fire.

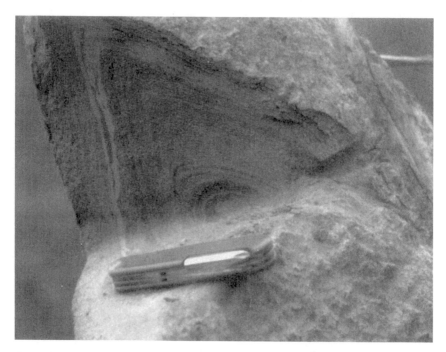

83 The eruption of Taal Volcano in the Philippines in 1965 generated pyroclastic surges that swept 2 to 3 kilometers across the southern part of Volcano Island. The sides of the trees facing the vent were plastered with layers of mud from the surges. The temperature of the hot surges that deposited these mud layers must have been lower than 100°C for the wet mud to stick to the trees.

> . . . After the explosion, not one living being was seen on land. Only 25 of those on the Roraima, out of 68, were left after the first flash. ("The Destruction of the Roraima," *Frank Leslie's Popular Monthly*, July 1902)

Though it must have seemed so, Mont Pelée was not blown to pieces. Although part of the old crater rim was destroyed, the flanks of the mountain were unchanged except for rapid erosion and deposition in the valley that directed the glowing avalanche from the summit down to St. Pierre (Figure 85). The relatively small deposits from this pyroclastic flow consisted of dense blocks of volcanic rock from the growing lava dome mixed with ash.

84 Large *nuée ardente* eruption from Mont Pelée, several months after a similar cloud of hot gases and volcanic ash destroyed St. Pierre and killed 28,000 people within minutes. (Photograph by A. Lacroix in 1902, courtesy of the American Museum of Natural History.)

85 St. Pierre was totally destroyed by the *nuée ardente* that swept down from Mont Pelée on May 8, 1902. Only two people in the city survived. (Photograph by Underwood and Underwood from the Library of Congress.)

A study of the explosive debris around a volcano tells us something about the nature of its previous eruptions and thus helps us predict the nature of possible future eruptions. For example, at Pompeii, the Roman city buried in A.D. 79 in pyroclastic deposits from Mount Vesuvius, the early stages of the eruption produced distinct airfall pumice deposits. Later accumulations of ash and pumice lumps indicate alternation of deposits from both pyroclastic surges and airfalls. This sequence helps explain the relatively small number of bodies found buried at Pompeii in proportion to the size of the city. Although most people apparently escaped during the early rain of pumice fragments, stragglers were smothered and buried by the pyroclastic surges that followed (Figure 86).

Because many newspaper accounts of explosive volcanic eruptions still incorrectly refer to pyroclastic flows as lava flows, the general nature

86 The cast of a dog buried in pyroclastic deposits at Pompeii. During the excavation of the city, holes were found with skeletal remains inside. By carefully injecting the hole with plaster, a cast of the corpse can be formed. About 2000 human victims have been unearthed at Pompeii. (Alinari/Editorial Photocolor Archives, Inc.)

of each is probably more important than their details. A *pyroclastic flow* is a heavier-than-air mixture of hot gases and hot volcanic fragments whose loose packing and internal motions keep the mixture fluidized. It flows rapidly like a low-viscosity fluid, but after losing speed and fluidization, it dumps its fragments into pyroclastic deposits. A moving *lava flow*, in contrast, consists of viscous molten rock. It moves much more slowly than a pyroclastic flow, and as it cools and hardens it becomes solid rock.

Oceanic-ridge volcanoes and Hawaiian volcanoes erupt mainly lava flows. If we count the unseen flows deep beneath the sea, lava becomes the major product of the Earth's volcanoes. Shallow submarine eruptions like Surtsey can build an island of explosive volcanic ash and bombs, but the great volume of magma from deeper eruptions goes into streams of underwater lava.

Individual lava flows are tonguelike in shape, much longer than they are wide. On the island of Hawaii, a typical lava flow might be 10 kilometers long, 200 meters wide, and 3 meters thick, but there is so much variation that any standard flow dimensions would be misleading. Each flow covers only a small fraction of the island with a finger of lava. This finger forms a low ridge, and later flows run either beside or between ridges formed by earlier eruptions. The process is like covering a jug with candle drippings; it takes hundreds of wax flows to build up a covering layer. The volcanic pile above sea level on the island of Hawaii is the accumulation of literally hundreds of thousands of lava flows.

A lava flow is hypnotic to watch. There is often a central river of orange-red molten rock 5 to 10 meters across, flowing at speeds of 5 to 50 kilometers per hour, depending on the slope. The flow then oozes out on all sides from this central stream and forms slowly advancing dark lobes of cooling lava rubble riding on the molten but unseen core of the flow. This rubble of lava blocks tumbles down the steep front of the advancing flow, sometimes showing a glimpse of the glowing interior, and is slowly overridden by the advancing flow. The growing edges and fronts of the flow look like giant slow-motion bulldozer treads moving out, down, and under as the mass of the flow spreads forward. The perimeter of the dark, growing flow is much larger than the central river of glowing, fast-flowing lava (Figure 87).

In Hawaii, the kind of lava flow just described is called an *a'a flow*, pronounced "ah ah." Its major characteristic is the rubble of broken lava blocks on its surface. Adventurers with thick boots have walked on these rubble surfaces while the flows are slowly moving, but a'a flows are hard on boots even after they have stopped and cooled.

Another kind of Hawaiian lava flow is called *pahoehoe*, pronounced "pa hoy hoy." These flows are generally thinner than the a'a flows and form a smooth to wrinkled surface of solid rock (see Figure 79). Old Hawaiian foot trails favor the ancient pahoehoe flows, and for good reason. Bare feet will last for many more kilometers on their smoother, more solid surfaces (Figure 88).

Sometimes the main current within a pahoehoe flow crusts over and forms a tunnel filled with a fast-moving stream of lava. As the eruption wanes, the lava in this tunnel drains out, leaving an empty cave known as a *lava tube* inside the cooled flow. These tubes are 1 to 10 meters in diameter, and some can be followed underground for many kilometers.

The barren, smooth surfaces of pahoehoe flows and the rubble that forms a'a flows are so distinctive that the Icelanders also have two separate words—*helluhraun* for pahoehoe and *apalhraun* for a'a—to describe these two kinds of flows where they occur in Iceland. The largest lava flow in recorded history was an a'a (apalhraun) flow that issued from a 25-kilometer-long fissure in Iceland between June and November 1783. Named the Laki flow, it covered more than 500 square kilometers with lava, completely filling two deep river valleys in the process. The volume of lava produced by the Laki fissure during the several months of erup-

87 Flowing basaltic lavas, especially a'a flows, often have central channels that move at several kilometers per hour. These channels are red-hot (in this photograph, light gray), in contrast to the partly cooled, black margins of the flow, which advance more slowly, at rates of several meters per hour. (Photograph of the 1977 eruption of Piton de la Fournaise on Réunion Island by Pierre Vincent.)

88   Lava flows of a'a (left) and pahoehoe both issued from Kilauea Volcano in Hawaii in 1973. The dark, rough a'a is about 3 to 4 meters thick; it was covering the smoother, glistening pahoehoe when the flow stopped.

tion was 12 cubic kilometers, enough to fill Yosemite Valley to a depth of 300 meters (Figure 89).

The texture of lava and pyroclastic rocks is largely controlled by the number and size of the gas bubble holes they contain. *Pumice*, being mostly holes, is one extreme. This frozen glass froth is so light that pieces of it will float on water. Dense, solid lava without holes is the other extreme, with most volcanic rocks being somewhere in between (Figure 90).

Why do some volcanoes, like Mount St. Helens, explode, generating pyroclastic deposits, while others, like Kilauea, produce relatively quiet eruptions of lava? Several factors account for this difference, including the gas content and viscosity of the magma, the number of sites where gas bubbles can grow, and the suddenness of the pressure release.

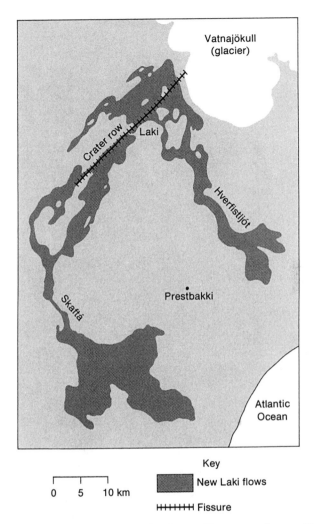

89 The Laki flow in south central Iceland. A 1783 eruption from a 25-kilometer-long fissure produced 565 square kilometers (12 cubic kilometers) of basaltic lava, a world record. (Map data from Sigurdur Thorarinsson.)

The gas content of magma is quite variable, ranging from less than 1 percent to more than 5 percent. At Mount St. Helens in 1980, the dissolved gases—mainly water and carbon dioxide—were estimated to be about 6 percent, by weight, of the magma; at Kilauea, the gas content of the magma amounts to about 1 percent. For this reason alone, Mount St. Helens had six times the explosive potential of Kilauea.

Mount St. Helens's magma was much more viscous than Kilauea's and confined the escaping gases in clots of magma until the pressures

90   Basaltic lava is often filled with gas bubble holes formed when steam and other gases dissolved in the magma are released by the low pressures at the Earth's surface. The tops of flows usually have more trapped bubble holes (vesicles) than do the centers of flows. The small crystals are olivine, one of the first minerals to form as the magma cools. (Photograph at true scale by J. D. Griggs, U.S. Geological Survey.)

◀──

built to levels that explosively burst the clots into fragments. A firecracker analogy is useful here: gunpowder confined in a paper wrapper explodes when ignited, while gunpowder poured from a broken firecracker into a small, open pile only burns. Of course, no combustion is involved in a volcanic explosion, but the nature of the gas release is analogous.

The places in a liquid where bubbles begin to form and grow are called *nucleation sites*. In magma, these sites are often tiny crystals in the melt. The more nucleation sites that an erupting magma contains, the faster the gases can escape. Once small bubbles have formed, more gas will enter them, causing the bubbles to grow rapidly. Two analogies are useful to picture this process. Shake salt into a glass of beer and it will foam much more rapidly because you have added nucleation sites. Shake a can of soda pop before pulling the tab and it will be more explosive because you have shaken many small bubbles (nucleation sites) from the cap of carbon dioxide gas beneath the can lid into the liquid.

The suddenness with which the pressure surrounding a magma body is changed is also a very important factor. If magma rises slowly toward the Earth's surface (that is, toward a region of lower pressure) over months or years, the gases can escape slowly through cracks in the surrounding rocks. However, if the pressure changes suddenly, as it did at Mount St. Helens when the avalanche "uncorked" the shallow magma intrusion and superheated groundwater, the gases will expand explosively.

The 1980 eruption of Mount St. Helens had all four factors driving its violent explosion: the high gas content of the magma, the high viscosity of the magma, the many tiny crystals already present in the magma, and the sudden decompression caused by the avalanche sliding away. Kilauea, in contrast, has a low gas content and low-viscosity magma, fewer tiny crystals present as gas bubble nucleation sites, and a relatively slower release of pressure than at Mount St. Helens.

Before lava or pyroclastic fragments are erupted, their parent material is called *magma*. Magma is a melt of rock-forming elements often mixed with solid crystals and gas bubbles. The common elements in the melt are oxygen, silicon, aluminum, iron, calcium, magnesium, sodium, titanium, and potassium, all in varying amounts. It is common to report the amount of silicon and aluminum present in a melt in combination with oxygen. Silica, the compound of one silicon and two oxygen atoms (SiO_2), is the most abundant constituent; alumina (Al_2O_3) is second. The proportion of silica varies from about 45 to 78 percent of the total. Various types of magma and their resultant igneous rocks are classified by their silica content.

Geologists like to name rocks, and there are hundreds of varieties to identify. However, most volcanic rocks fall into a spectrum of compositions, with segments named basalt, andesite, dacite, or rhyolite, which have a silica content of about 45 to 54 percent, 54 to 62 percent, 62 to 70 percent, and 70 to 78 percent, respectively (Table 2). Rocks that contain more silica also contain more sodium and potassium and less iron, calcium, and magnesium. These last three elements, particularly the iron, form dark minerals, so that basalts are dark gray, almost black; andesites are medium gray; and dacites and rhyolites are usually light gray to tan. One important exception is obsidian, a dense, glassy form of rhyolite that was once used for arrowheads. Obsidian is nearly black. Although there is less iron in rhyolite than in other kinds of volcanic rocks, in obsidian it is so finely dispersed as tiny crystals of opaque iron oxides that it makes the glass look shiny black.

The chemistry of volcanic rocks is only part of the story of their complex composition. As magma cools, various minerals start to crystallize from the melt. Crystallization takes time. If basaltic magma cools rapidly, within seconds or minutes, the compounds within it will not have time to arrange themselves into minerals, and the result is a dark, opaque glass. If the magma cools more slowly, over days or years, the minerals will have time to form and grow. Viscosity also affects crystal growth; crystals form more slowly in high-viscosity magmas, and so some thick rhyolitic flows can be nearly all glass.

TABLE 2   Types of igneous rock

Range of SiO_2 content	Extrusive (erupted on surface)	Intrusive (solidified below surface)
45%–54%	Basalt	Gabbro
54%–62%	Andesite	Diorite
62%–70%	Dacite	Granodiorite
70%–78%	Rhyolite	Granite

The cooling of molten basalt in Kilauea Iki lava lake in Hawaii provides a good example of volcanic rock formation. The lava was erupted in 1959 at temperatures up to 1200°C and was nearly all liquid except for a small percentage of olivine crystals. Most of the water and other gases dissolved in the magma were vaporized and lost in the fume that was formed as the lava sprayed out of the vent in fountains and then ponded in an older crater. The surface of the lake cooled quickly and formed a glassy crust with many bubble holes.

A project was set up to drill a sequence of holes in the thickening lava crust and to study the cooling and crystallization of the basalt under the slowly changing and controlled conditions within the lava lake. During the drilling, large volumes of water were pumped down the holes to prevent overheating of the drill bits and core collectors, and temperatures were measured by electrical thermocouples lowered down the empty holes after the drilling had stopped and the holes had been reheated by the molten rock below.

The deeper molten lava cooled more slowly because of the insulating effect of the overlying crust, and crystals of various minerals began to appear. Olivine crystallized first; because this mineral crystallizes at temperatures from 1250°C to 1190°C, some crystals were present even at the time of eruption. Pyroxene formed next, at temperatures between 1190°C and 1180°C, and then plagioclase feldspar at 1170°C to 1160°C. By the time the temperature had dropped to 1065°C, the basalt was half liquid and half crystals of plagioclase, pyroxene, and olivine, and it had become hard enough that steel probes could not be pushed into the bottom of shallow drill holes. Magnetite began to crystallize at 1030°C, forming the black mineral grains that give the basalt much of its dark color. By 980°C the basalt was all crystals except for 5 to 10 percent glass. Although it was still red-hot, it had effectively become a solid.

During cooling and crystallization, the composition of each mineral that crystallizes is quite different from the overall composition of the original magma. This differentiation keeps changing the composition of the remaining melt fraction as the crystallization progresses (Figure 91). Its most evident effect is seen in the composition of the final interstitial glass by the time cooling has reached 980°C. This glass has a silica content of 60 percent, compared with 48 percent in the original melt.

The cooling of andesitic and rhyolitic magmas is equally complex, and other silicate minerals, including quartz, become important constituents of the volcanic rocks formed from these magmas. An important difference between rocks formed from a cooling silicate magma and ice formed by cooling water is the mixture of minerals in igneous rocks. Wa-

91 A photomicrograph of a very thin slice of Hawaiian basalt from Kilauea Iki lava lake. Drilling core holes into the slowly cooling lake provides samples of "quick-frozen" basalt in which the drilling water suddenly chills the slowly cooling melt and arrests the process of crystallization. The large crystal on the right is olivine, and the smaller, sticklike crystals are feldspar. The uniform gray area surrounding the silicate crystals is glass formed from the sudden cooling of the melt. (Photograph by the U.S. Geological Survey.)

ter freezes at 0°C and forms only one mineral, but magma freezes from 1250°C down to 700°C, and the final, solid rock contains two or three or more minerals.

The volumes of lava and pyroclastic deposits produced in individual recorded eruptions range from a few cubic meters up to 50 cubic kilometers (Table 3). On average, subduction-zone volcanoes produce about 1 cubic kilometer of new volcanic rock each year, composed mostly of pyroclastic deposits.

The volume of volcanic rock produced by oceanic-ridge volcanoes is largely hidden in the deep oceans, but it can be estimated indirectly. The spreading rates and sizes of the plates are reasonably well established, and these data allow an estimate of the new area of seafloor that must be created every year to fill in the cracks opening between the spreading plates. The estimate is about 2.5 square kilometers per year. Allowing 1.5 kilometers for the general thickness of basaltic flows that accumulate on the seafloor, the average amount of oceanic-ridge eruptive products is about 4 cubic kilometers per year.

TABLE 3 Volcanic Explosivity Index (VEI)

	0	1	2	3	4	5	6	7	8
General description	Non-explosive	Small	Moderate	Moderate-large	Large	Very large			
Volume of tephra (m³)	1×10^4	1×10^6	1×10^7	1×10^8	1×10^9	1×10^{10}	1×10^{11}	1×10^{12}	
Cloud column height (km) Above crater		<0.1	0.1–1	1–5					
Above sea level				3–15	10–25	>25 ⟶			
Qualitative description	"Gentle"	"Effusive" ⟵	— "Explosive" ⟶	⟵	"Cataclysmic" "Severe"	"Paroxysmal" "Violent"	"Colossal" "Terrific" ⟶		
Eruption type		⟵ Strombolian ⟶ ⟵ Hawaiian ⟶	⟵ Vulcanian ⟵	⟵ Plinian ⟶ Ultra-Plinian ⟶					
Duration (continuous blast)		<1 hr> ⟶ ⟵ 1–12 hrs ⟶			⟵ >12 hrs ⟶				
CAVW[1] Max explosivity	Lava flows ⟵ ⟵ Phreatic ⟶			Explosion or *nuée ardente* ⟶ - - - - ⟶					
Tropospheric injection	Negligible	Minor	Moderate	Substantial ⟶					
Stratospheric injection	None	None	None	Possible	Definite	Significant ⟶			
Eruptions[2]	699	845	3477	869	278	84	39	4	0

Source: Adapted from C. G. Newhall and S. Self, "The Volcanic Explosivity Index (VEI)," *Journal of Geophysical Research* 87 (1982), 1231–1238.

[1]CAVW refers to Catalog of Active Volcanoes of the World, International Association of Volcanology and Chemistry of the Earth's Interior, Parts I–XXII, Rome, Italy, 1951–1975.

[2]Number of eruptions of each magnitude in Tom Simkin and Lee Siebert, *Volcanoes of the World*, Smithsonian Institution Press, 1994.

Hawaiian volcanoes average about 0.1 cubic kilometer per year of basaltic lava flows, but the total for all hot-spot volcanoes is difficult to estimate because of the long intervals between eruptions in places like Yellowstone. A reasonable guess is an average of 0.5 cubic kilometer per year for the volcanic products of hot spots. The Earth's present volcanoes thus have a total output of about 5 to 6 cubic kilometers of new rock each year on average. Many years go by without a large volcanic eruption, and then a Pinatubo-type eruption comes along to catch up on the lag. In addition to the erupted rocks, shallow injections of magma intruded and crystallized beneath volcanoes also amount to large volumes, probably more than the erupted rocks.

Huge deposits of pyroclastic flows that cover thousands of square kilometers and are tens to hundreds of meters thick are found in Japan, New Zealand, Central America, the western United States, and many other volcanic regions of the world. Some of these deposits give every indication that they were poured out in single, enormous eruptions that would dwarf Pinatubo and Katmai. The volume of these deposits is on the order of 300 to 3000 cubic kilometers, compared with the 30 cubic kilometers from Katmai. This observation raises some fundamental questions. Was prehistoric volcanism—say, a million years ago—greater than it is now? Or is the recorded history of volcanic eruptions so short—2000 years in Europe and about 200 years in the United States—that the data we have cover too brief a time to be representative? We think the latter is true. Pinatubo and Katmai are probably only small samples of what nature can deliver in the way of volcanic cataclysms.

Links

Aircraft Volcanic Ash Advisories
http://www.ssd.noaa.gov/VAAC/messages.html

Big List of Volcano Links
http://vulcan.wr.usgs.gov/Servers/earth_servers.html

Crater Lake, Oregon
http://geopubs.wr.usgs.gov/fact-sheet/fs092-02/

Hawaiian versus Mt. St. Helens Eruptions
http://vulcan.wr.usgs.gov/Photo/Pictograms/hawaii_cascades_erupt.html

Hawaiian Volcano Observatory
http://hvo.wr.usgs.gov/

Pyroclastic Flows
http://vulcan.wr.usgs.gov/Photo/Pictograms/pyroclastic_flows.html

Volcanic Ash—Danger to Aircraft
http://geopubs.wr.usgs.gov/fact-sheet/fs030-97/

Volcanic Ash, U.S. Geological Survey
http://geopubs.wr.usgs.gov/fact-sheet/fs027-00/

Volcano World
http://volcano.und.edu/

Weekly Volcanic Activity
http://www.volcano.si.edu/gvp/reports/usgs/index.cfm

Sources

Cas, R. A. F., and J. V. Wright. *Volcanic Successions*. London: Allen & Unwin, 1987.

Fisher, R. V., and H.-U. Schmincke. *Pyroclastic Rocks*. Berlin: Springer-Verlag, 1984.

Francis, Peter. *Volcanoes: A Planetary Perspective*. New York: Oxford University Press, 1993.

Hibbard, M. J. *Petrography to Petrogenesis*. Englewood Cliffs, NJ: Prentice-Hall, 1995.

Schmincke, Hans-Ulrich. *Volcanism*. Berlin: Springer, 2004.

Sigurdsson, Haraldur, ed. *Encyclopedia of Volcanoes*. San Diego, CA: Academic Press, 2000.

Simkin, Tom. "Terrestrial Volcanism in Space and Time." *Annual Review of Earth and Planetary Sciences* 21 (1993): 427–452.

Steingrimsson, Jon. *Fires of the Earth: The Laki Eruption 1783–1784*. Reykjavík: University of Iceland Press, 1998. (A first-person account.)

CHAPTER 10

Lahars and Avalanches

92 The summit of Colombia's Nevado del Ruiz Volcano, 5389 meters high, is covered by a 17-square-kilometer ice cap. A small pyroclastic eruption on November 13, 1985, melted about 5 to 10 percent of the ice cover and caused mudflows in the valleys draining the mountain's steep flanks. (Photograph by Robert Tilling, U.S. Geological Survey.)

> *Muddy: not clear, containing sediment, obscure in meaning.*
> —Webster's Dictionary

In stream valleys near explosive volcanoes, it is common to see terraces and banks composed of poorly sorted gravel in a matrix of sand, silt, and clay. These are deposits from *lahars*, the Indonesian name for volcanic debris flows—more commonly known as *mudflows*. Lahars can occur when heavy rains wash down loose ash from earlier eruptions or when an earthquake dislodges a steep slope of clay and rocks that have been deeply weathered by volcanic gas vents and hot springs. At other times and places, lahars can result when an eruption ejects a crater lake or rapidly melts the snow and ice on a volcano's high summit. As these huge floods of water surge down a volcano's flanks, they pick up volcanic ash and soil and quickly thicken into lahars. The surging mud in turn picks up larger rocks and boulders, and the entire mass thunders down a valley like a torrent of wet concrete.

The 1985 tragedy at Nevado del Ruiz, a 5389-meter-high volcano in the Andes of Colombia, was entirely the result of a lahar. Although Ruiz is only 5 degrees north of the equator, its high summit is covered with a large snow and ice field (Figure 92). In 1845 an eruption or earthquake triggered a lahar from Ruiz that killed about a thousand people, and the new town of Armero was built on top of it. In November 1985, a relatively small eruption of hot pumice and ash jetted from the crater, landed on the ice cap, and caused sudden melting. Huge floods of water mixed with ash poured off the summit, most of it rushing down the tributaries of the Lagunillas River, picking up soil, rocks, and trees as it swept along. In places these floods—now debris flows—scoured the canyon walls to heights of 80 meters above the normal stream levels.

The lahar roared down the Lagunillas Valley at speeds of 50 to 60 kilometers per hour. The town of Armero was located where this narrow canyon meets the wider, open valley of the Magdalena River. At the canyon's mouth, the torrents spilled out on Armero in a debris flow as much as 30 meters high, which quickly spread out into dense flows 3 to 5 meters deep. Armero was swept away, and more than 20,000 people died. The deposit of dried mud and debris left behind is about 1 meter thick and covers an area of 40 square kilometers. Boulders as large as 10 meters lie scattered where the town once stood (Figure 93).

93 The town of Armero, Colombia, 50 kilometers east of and 5 kilometers below the summit of Nevado del Ruiz Volcano, was destroyed by a massive mudflow that swept down the Lagunillas River from the November 13, 1985, eruption. About 22,000 people in Armero were killed.

When we wrote the first edition of this book, it was thought that the character of a mudflow deposit provided little or no information about the type of event that caused it. But now the eruption of Mount St. Helens (see Chapter 4) and the studies of Kevin Scott of the U.S. Geological Survey and his colleagues have changed all that. As in most detailed studies, it's important to get the terms straight before coming to conclusions.

Scott distinguishes two types of lahars: *cohesive debris flows* containing more than 3 to 5 percent clay (particles smaller than 4 microns) and *noncohesive debris flows* containing smaller amounts of clay. A cohesive debris flow maintains its "wet concrete" character throughout its course, but a noncohesive flow begins as a flood, thickens into a hyperconcentrated flow (one with 20 to 60 percent sediment by volume), continues thickening into a true lahar at more than 60 percent sediment, and then reverses this sequence as it loses velocity. The higher clay content in a cohesive lahar apparently holds the flow intact, in contrast to the noncohesive lahar, which picks up or loses its sediment load depending on its velocity. The character of the deposits from these two types of lahars thereby provides some clues to their origin.

For example, the large lahar that swept down the North Toutle River during and right after the May 18, 1980, eruption of Mount St. Helens was of the cohesive variety. Its water and mud were derived in part from compaction and melting of ice in the huge debris avalanche, and possibly from the giant waves of water displaced by the part of the avalanche that plunged into Spirit Lake. The lahar from Ruiz that buried Armero was of the noncohesive variety. The enormous, 3-cubic-kilometer Osceola Mudflow from Mount Rainier that reached Puget Sound 5700 years ago was a cohesive lahar generated by a major summit collapse and debris avalanche (Figure 94).

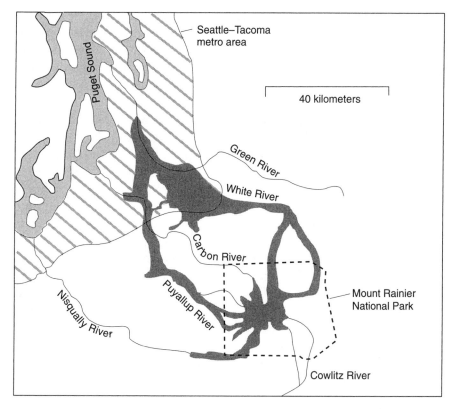

94 Prehistoric mudflows from Mount Rainier, an active Cascade volcano, have covered part of what is now the Seattle–Tacoma metropolitan area. The giant Oceola Mudflow swept down the White River about 5700 years ago, covering more than 500 square kilometers. At nearly the same time, the Paradise Lahar rushed down the Nisqually River. As recently as 500 to 600 years ago, the Electron mudflow covered much of the lowlands along the Puyallup River. (After U.S. Geodynamics Committee, *Mount Rainier: Active Cascade Volcano*, National Academy Press, 1994, p. 25.)

Although huge cohesive lahars occur less frequently than do the smaller, noncohesive lahars, they can travel long distances from their sources. A cohesive lahar can be triggered by a major earthquake as well as by a volcanic eruption.

A *glacial burst* is a unique type of volcanic flood that sometimes occurs in Iceland. A volcanic eruption beneath a thick ice cap can melt a huge volume of ice, and the meltwater accumulates in a lake or subglacial reservoir that is trapped under the ice. Increasing pressure from the reservoir's rising level eventually forces the water to move beneath the ice cap and escape through one of the river valleys that drain the glacier. These huge floods are called *jokulhlaups*—Icelandic for "glacial bursts."

A recent jokulhlaup occurred on November 5, 1996, five weeks after an eruption near Grímsvötn Volcano, a caldera under the Vatnajökull ice cap. The ice-covered lake that accumulated in the caldera was estimated to contain some 3 to 4 cubic kilometers of meltwater before it forced its way beneath the glacier. Grímsvötn Caldera is nearly buried by the ice cap, but some of its rocky rim protrudes through the ice—appearing like an atoll rising above the ocean. A seismograph that had been placed on this rim provided the first warning of the flood. As the water surged beneath the ice, it generated seismic vibrations that started ten hours before the flood burst out. This warning enabled the authorities to close the road that crosses the plain between the ice cap and the sea, and no one was killed. The peak discharge of the flood was 45,000 cubic meters per second, nearly three times the average flow volume of the Mississippi River. A bridge and 10 kilometers of the road disappeared, while two other bridges and 10 more kilometers of the road were badly damaged. Giant blocks of ice covered the plain as the flood subsided.

An *avalanche* is a large mass of snow, ice, soil, or rock—or a mixture of these materials—that falls or slides rapidly down a cliff or steep slope. An avalanche composed of a mixture of materials is called a *debris avalanche*. Large avalanches from the summit and steep slopes of a volcano pose major hazards that may or may not be related to an eruption. A debris avalanche can be triggered by heavy rainfall or a strong earthquake without a volcanic eruption. At Mount St. Helens, however, the great avalanche and the eruption were closely connected. The injection of a shallow body of magma caused a bulge on the already steep north slope of the volcano. This oversteepened slope was shaken loose by a moderate earthquake into the great avalanche, which in turn ripped open the mountain and unleashed the explosive eruption. These close connections between large avalanches and eruptions may be more common than previously thought.

Horseshoe-shaped craters that have formed where a combination of avalanche and explosion has torn out a whole section of a steep volcanic cone are now being recognized on several active volcanoes. With knowledge gained from the Mount St. Helens avalanche and eruption, geologists have finally solved the mystery of the extensive area of hills and mounds of broken volcanic rock northwest of Mount Shasta in California. They now believe that between 300,000 and 380,000 years ago a gigantic avalanche from an ancestral peak of Mount Shasta rushed down the Shasta River valley for 43 kilometers. The hummocky debris-avalanche deposit covers 450 square kilometers and has a volume of 45 cubic kilometers. It is the largest subaerial landslide known to have occurred anywhere in the world during the last million years (Figure 95).

Because of their steep slopes and high snow-and-ice-covered summits, avalanches and lahars are more common on stratovolcanoes than on the more gently sloping and rounded shield volcanoes. But such assumptions are sometimes shaken by new discoveries. For instance, undersea surveys around the shield volcanoes of the Hawaiian Islands dur-

95 This landscape of hills north of California's Mount Shasta (in the background) is part of a gigantic avalanche deposit that slid from the volcano sometime between 300,000 and 380,000 years ago. This debris avalanche traveled more than 40 kilometers beyond the base of the mountain and contains 45 cubic kilometers of broken volcanic rock. (Photograph by C. Dan Miller, U.S. Geological Survey.)

ing the past decade have revealed some submarine landslides that make the Shasta avalanche look small.

In 1983 the United States declared sovereignty over seabed resources extending for 200 nautical miles off its shores, and the USGS was given the task of mapping and evaluating this new area. One of the most useful tools in this survey was side-scanning sonar towed by surface vessels that could image a swath of sea bottom as much as 50 kilometers wide. The seabed off the shores of the Hawaiian Islands from Midway Island to the island of Hawaii was mapped between 1986 and 1991. The astounding outcome of this survey was the discovery of at least 68 giant underwater landslides more than 20 kilometers long. Streams and waves had been thought to be the great destroyers of the Hawaiian volcanoes—instead, it turns out to be gravity.

Two dominant types of Hawaiian submarine landslides have been identified: slumps and debris avalanches. A *slump* moves on an average slope of more than 3 degrees and involves a large mass of a volcano that slides with little internal disruption. A 100-kilometer-wide slump block can extend from a subaerial volcanic rift zone to its submarine terminus more than 20 kilometers at sea and be as much as 10 kilometers thick (Figure 96). According to studies of the southeast flank of Kilauea Volcano, the movement of a slump block is apparently either a slow creep of about 10 centimeters per year or a sudden slip of a few meters that can generate a major earthquake like the magnitude 7.2 quake that occurred there in 1975.

In contrast, a giant *submarine debris avalanche* slides on an average slope of less than 3 degrees and can move slightly upslope. It moves rapidly and breaks up some of the avalanche block into fragments. An avalanche is thinner and longer than a slump, and its run-out is marked by hummocky terrain with broken blocks as much as 1 to 10 kilometers in diameter. Some avalanches, such as Alika 1 and 2 off the west coast of the Big Island, have left well-developed escarpments on land where they slid away. These steep slopes have been draped with younger lava flows from Mauna Loa Volcano. The largest debris avalanches have volumes on the order of 5000 cubic kilometers—the largest avalanches known on Earth.

The Hilina slump on the southeast coast of the Big Island is still active, but the youngest avalanche (Alika 2) has an inferred age of about 100,000 years, based on tsunami deposits on the Hawaiian island of Lanai. Can a slump turn into an avalanche? Some apparently have, where the toe of a slump has become oversteepened. Could the active Hilina slump suddenly slide away? Perhaps. Geologists have speculated that a very large Hilina earthquake exceeding magnitude 8 might trigger a catastrophic failure of the southeast flank of Kilauea Volcano. But *perhaps* and *maybe* are favorite words of geologists.

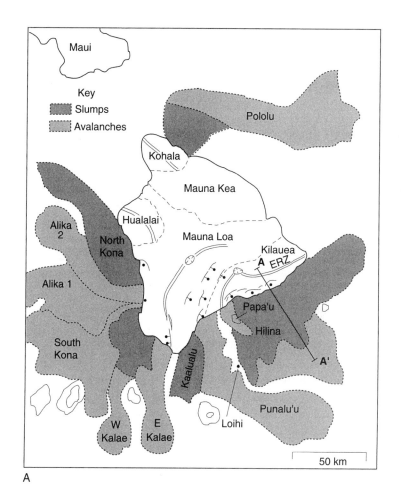

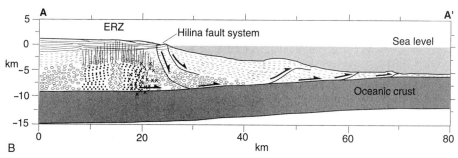

96 Giant submarine landslides on the flanks of the Hawaiian Islands are among the largest on Earth. Some of them are longer than 200 kilometers, and some have volumes of thousands of cubic kilometers. Two types are seen on the sonar surveys. Large slumps move by slow creep and intermittent small displacements during large earthquakes. The Hilina slump on the southeast flank of Kilauea Volcano is an active example. Huge debris avalanches move long distances at high speeds and are caused by catastrophic failure. These avalanches, like Alika 2 on the west coast of Hawaii, are thought to have generated enormous tsunamis. Fortunately, their recurrence interval is in the range of hundreds of thousands of years. (After James Moore, William Normark, and Robin Holcomb, Palo Alto, CA: *Annual Reviews of Earth and Planetary Science*, 1994, p. 128.)

One interpretation of the geologic record is that two giant tsunamis left their marks on the islands of Lanai and Molokai 100,000 and 200,000 years ago, respectively. These huge waves are thought to be related to the disturbance of the ocean by large submarine avalanches. The Lanai tsunami appears to have washed blocks of coral into extensive surface deposits as high as 325 meters above current sea level, and the Molokai deposits are as high as 70 meters above current sea level. In 1975 a tsunami with a run-up height of a few meters washed up on the southeast coast of the Big Island soon after the magnitude 7.2 earthquake on the Hilina slump. In 2003, some geologists examining the high-elevation coral deposits on Lanai concluded that they were more likely formed at sea level and that the entire island had been slowly uplifted. This has sparked a lively debate and, as the saying goes, "the jury is still out." There seems to be no disagreement, however, about the evidence of giant submarine landslides from the flanks of the Hawaiian Islands.

Like many geologic processes, small landslides occur frequently and large events more rarely. Is a catastrophe with an average recurrence time of hundreds of thousands of years worth worrying about? For a person whose average life span is less than one-thousandth of this recurrence time, the answer is certainly no; there are far more immediate dangers to worry about. What is fascinating is not the danger of these giant underwater landslides, but the fact that they were virtually unknown only a few years ago.

Other surveys are already turning up similar landslides at several volcanic islands, including Réunion Island in the Indian Ocean, Stromboli and Mount Etna in the Mediterranean, Tristan da Cunha and the Canary Islands in the Atlantic, and the Marquesas Islands in the Pacific. What started as a survey to evaluate the resources on the seabed near Hawaii revealed an entirely new stage in the life cycle of oceanic volcanoes.

Links

Avalanches Theme Page
  http://www.cln.org/themes/avalanches.html

Hawaiian Submarine Landslides
  http://volcano.und.nodak.edu/vwdocs/vwlessons/evolution/part7.html

Hysteresis and Avalanches
  http://www.lassp.cornell.edu/sethna/hysteresis/hysteresis.html

Lahars, Michigan Tech University
  http://www.geo.mtu.edu/volcanoes/hazards/primer/lahar.html

Lahars of Pinatubo
  http://wrgis.wr.usgs.gov/fact-sheet/fs114-97/

Mount Hood
http://geopubs.wr.usgs.gov/fact-sheet/fs060-00/

Mount Rainier
http://geopubs.wr.usgs.gov/fact-sheet/fs034-02/

Mudflows, Debris Flows and Lahars
http://vulcan.wr.usgs.gov/Glossary/Lahars/framework.html

Nevado del Ruiz
http://vulcan.wr.usgs.gov/Volcanoes/Colombia/Ruiz/framework.html

Submarine Volcanism
http://www.mbari.org/volcanism/Hawaii/HR-Landslides.htm

U.S. Geological Survey, Landslides
http://landslides.usgs.gov/

U.S. Geological Survey's Volcanic Hazards Program
http://volcanoes.usgs.gov/

Sources

Carlowicz, M. "Glacial Burst Overwhelms Skeidara." *Eos* (*Transactions, American Geophysical Union*), November 19, 1996.

Crandell, D. R. "Gigantic Debris Avalanche of Pleistocene Age from Ancestral Mount Shasta Volcano, California, and Debris-Avalanche Hazard Zonation." U.S. Geological Survey Bulletin 1861, 1989.

Moore, James, W. B. Bryan, and K. R. Ludwig. "Chaotic Deposition by a Giant Wave, Molokai, Hawaii." *Geological Society of America Bulletin* 106 (1994): 962–967.

Moore, James, W. R. Normark, and R. T. Holcomb. "Giant Hawaiian Landslides." *Annual Review of Earth and Planetary Sciences* 22 (1994): 119–144.

Moore, James, W. R. Normark, and R. T. Holcomb. "Giant Hawaiian Underwater Landslides." *Science* 264 (1994): 46–47.

Morgan, J. K., G. F. Moore, and D. A. Clague. "Slope Failure and Volcanic Spreading along the Submarine South Flank of Kilauea Volcano, Hawaii." *Journal of Geophysical Research* 108 (2003): 2415.

Scott, Kevin. *Origins, Behavior, and Sedimentology of Lahars and Lahar-Runout Flows in the Toutle–Cowlitz River System.* U.S. Geological Survey Professional Paper 1447-A, 1988.

Scott, K. M., J. W. Vallance, and P. T. Pringle. *Sedimentology, Behavior, and Hazards of Debris Flows at Mount Rainier, Washington.* U.S. Geological Survey Professional Paper 1547, 1995.

Takahashi, E., P. W. Lipman, M. O. Garcia, J. Naka, and S. Aramaki. *Hawaiian Volcanoes: Deep Underwater Perspectives.* Geophysical Monograph 128. Washington, DC: American Geophysical Union, 2002. 418 pp.

CHAPTER 11

Cones, Craters, and Calderas

97 Pavlof and Pavlof's Sister, stratovolcanoes in the Aleutian Islands, Alaska. (Photograph by Katia Krafft.)

Mountains come in different styles. Some, like the Himalayas, have been thrust upward where the crust of the continents is telescoped together by powerful compressive forces. Others, like the ranges between the Rockies and the Sierra Nevada, were formed by regional uplift and the downfaulting of keystone blocks into basins between the ranges as the crust pulls apart under tension. Still others, such as the Appalachians, resulted from regional uplift and the subsequent etching by erosion of hard and soft rock formations—folded by earlier compression—to form ridges and valleys.

Volcanoes have a style all their own. They are built upon the landscape by outpourings of new rock. Their forms change rapidly, some even during a person's lifetime. In 1975 in the Tolbachik region of Kamchatka, a new cinder cone grew to a height of 600 meters in just four months. The known record for rapid growth goes to Vulcan, a volcano on the rim of Rabaul Caldera in Papua New Guinea. The 1937 eruption lasted only four days, and most of Vulcan's 230-meter-high cone was built during the first twelve hours.

Geologists learn that the hills are not everlasting, that erosion slowly strips away the mountains and fills the intervening basins with their debris. It is usually taken for granted these days that river valleys were carved out by the streams they contain. But move to an active volcanic terrain and all the rules change. The mountains are growing, new rocks are created, and the valleys are often the spaces between lava flows. In other words, some valleys are formed not by erosion, but by lack of construction. Even experienced geologists must remind themselves that they are in a strange land where the rules are often reversed.

Volcanoes create some of the most beautiful and unusual landscapes on Earth. Their power and fire command respect and are reminders of the intense vitality of our Earth. *Kazan*, the Japanese word for volcano, means "fire mountain"; our English word *volcano* comes from Vulcano, the name of an Italian island whose volcanic peak was considered the forge of Vulcan, the Roman god of weapons.

The shapes of volcanic mountains range from the perfect cones of Mount Fuji in Japan and Mayon in the Philippines to flat lava plateaus in Iceland that barely meet the definition of a mountain. The volcanic vents on these mountains range from vast circular basins called *calderas*, many kilometers in diameter, to tiny *craters* a few meters across (Figure 98). Several factors are at work in producing these great variations in form.

The shape of the vent from which the volcanic products escape to the surface is of prime importance. Often the vent is a long, nearly vertical crack in the ground, several hundreds to thousands of meters long and deep and only a few meters wide. Magma rises through this fracture and erupts along the linear vent at the surface, sometimes welling out in wide streams of lava and sometimes in a line of lava fountains. The Laki eruption in Iceland and the beginning of the Pu'u O'o eruption in Hawaii are examples of eruptions from linear vents.

A specific location along a linear vent usually becomes the principal source of volcanic emissions as an eruption progresses, forming a central

98 Summit of Cotopaxi Volcano, Ecuador. The nested craters resulted from explosive eruptions of different intensities. Snow and ice dominate the summit of this 5900-meter peak despite its volcanic activity and its location less than 1 degree south of the equator. (Photograph by G. E. Lewis, U.S. Geological Survey.)

vent from which a volcanic mountain begins to grow. If a volcano is eroded down to its roots, the hardened magma in the exposed feeder cracks, which has greater resistance to erosion than the rock surrounding it, may remain as long, low ridges. These features are called *volcanic dikes*. The eroded remnant of a steep, pipelike conduit that fed a central vent is called a *volcanic neck* (Figure 99).

The viscosity of the magma as it erupts also has a profound effect on the shape of a volcano. Very viscous lava squeezes up into a steep-sided plug over its vent, called a *lava dome*. Lassen Peak, a 27,000-year-old volcano in northern California, is one of the world's largest known lava domes (Figure 100). At this writing (March 2005), a new lobe of the lava dome in the crater of Mount St. Helens is erupting.

Solid fragments thrown from a vent form a pile of debris called a *cinder cone* around or downwind from the crater (see Figure 52). These cones have straight sides with slopes of about 30 degrees, the angle of re-

99 Ship Rock in New Mexico is a volcanic neck with radiating dikes. Erosion has stripped away the volcanic cone, leaving only the hard skeletal remains. Ship Rock is 450 meters high, and the dikes form great walls stretching across the desert. (Photograph from John S. Shelton, *Geology Illustrated*, p. 15. Copyright © 1966, W. H. Freeman and Company.)

100   The lava dome that forms Lassen Peak in California was erupted about 27,000 years ago. The 2-cubic-kilometer volume of the dome puts it among the world's largest. Lassen last erupted between 1914 and 1917, with the principal explosive activity occurring in May 1915.

pose above which the cinders slide until a stable angle is established. Very fluid lavas, on the other hand, flow long distances on gentle slopes, forming lava plateaus or low-sloping volcanic piles called *shield volcanoes*. Since the composition of a lava is generally related to its viscosity — basalt is more fluid and rhyolite more viscous—the shape of a volcano is an important clue to its composition (Figure 101).

Many subduction volcanoes begin erupting with an explosion of volcanic ash that is followed by flows of lava. The alternation of ash and lava flows forms the steep concave slopes of the classic volcanic cone. The scientific name for this type of structure is *stratovolcano*, or *composite cone*, because of the layering of pyroclastic deposits and lava flows (Figures 97, 102, 103, and 104).

The surface environment of the volcanic vent also affects the shape of the resulting volcano. Submarine volcanoes provide the best examples of the effects of such environmental factors. As explained in Chapter 3, deep submarine volcanoes are less explosive because the water pressure is so great that steam cannot form and expand. But because water cools the lava faster than air would, the lava piles are generally steeper than

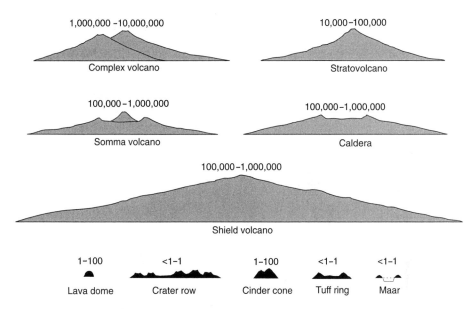

101 The relative sizes, shapes, and life spans of various types of volcanoes. The shaded profiles of volcanoes are vertically exaggerated two times, and the black profiles of craters and calderas are exaggerated four times. Their approximate life spans of active volcanism are indicated in years. (Adapted from Tom Simkin and Lee Siebert, *Volcanoes of the World*, Smithsonian Institution Press, 1994.)

those same lavas would be if they erupted above water. Shallow submarine eruptions usually are explosive, even when they involve basaltic magma that would be poured out on land in relatively quiet fountains and flows.

Tuff rings and maars are closely related volcanic landforms that are built by steam explosions caused by rising magma entering groundwater at a shallow depth. A *maar* is a flat-floored explosion crater with a low rim of debris derived largely from the country rock of the excavated crater rather than new volcanic material. The crater often fills with groundwater, forming a circular lake. The Eifel craters in Germany are examples of maars. A *tuff ring* is built by multiple pyroclastic surges that build a ring of debris around an explosion crater. The deposits that build the ring are a combination of the excavated country rock and new volcanic material. Diamond Head in Honolulu, Hawaii, is a classic example of a tuff ring (Figure 105).

Volcanic eruptions beneath thick piles of glacial ice show many of the same features as submarine eruptions: deep outpourings of pillow lava, followed by shallow-water explosions of volcanic debris, and capped by gently sloping flows of lava that reach above the glacial surface. The

102   Initial stages of an explosion cloud forming on Lassen Peak in California. This volcano erupted from 1914 to 1917. (Photograph by B. F. Loomis, June 14, 1914, courtesy of the National Park Service.)

103   Mayon Volcano, in the Philippines, erupting. The rising ash column and descending pyroclastic flow partly conceal the perfect symmetry of Mayon's stratovolcano cone. (Photograph by Dainty Studio, Daraga, Albay, Philippines, April 30, 1968.)

104 Santiago Crater of Masaya Volcano in Nicaragua reveals the composite inner structure of a stratovolcano. The light layers are lava flows, and the dark layers are ashfalls. The thick dark layer at the top is part of a cinder cone whose summit is just to the left of this photograph.

table-mountain volcanoes of Iceland are thought to have originated in this way (Figure 106). The glacial ice caps are now largely gone, but the steep shoulders of the table mountains mark their former thickness.

Other factors, such as the volume of erupted material and the length of time between eruptions, also influence the shape of a volcano. Small pyroclastic flow deposits fill valleys, while huge pyroclastic flow deposits

105 Diamond Head, the prominent landmark east of Waikiki Beach in Honolulu, looks quite different when seen from the air. This tuff ring was built by a late phase of Hawaiian volcanism, some 3 million years after the main eruptions that created the island of Oahu. Pieces of coral in the tephra indicate that steam explosions generated by the interaction of erupting lava and seawater occurred in or below the shallow reefs that fringe the island.

106 Herdubreid, a table mountain in Iceland. Pillow lava on the steep flanks covered by gently sloping flows on the summit indicates that this mountain was formed by eruptions beneath a glacial ice cap. The elevation to the edge of the table marks the thickness of the ice. (Photograph by Gudmundur Sigvaldason.)

163

create *ignimbrite plateaus* that bury all the underlying topography. Los Alamos, New Mexico, is located on such a volcanic plateau, formed about 1 million years ago (Figure 107). Ignimbrite plateaus are common in many volcanic regions of the world, but fortunately this type of eruption has not taken place in recorded history. The great size and speed of such a pyroclastic flow would be cataclysmic.

The craters on volcanoes also are created in many different ways. Funnel-shaped craters usually result from an initial enlargement of the vent by volcanic explosions, followed by the collapse of loose debris back into the crater at the end of the eruption. After gas escapes from magma, the reduction in volume sometimes allows the magma to drain back down the vent, forming a deep cylindrical crater on the volcano's summit. The escape of magma from a vent on a volcano's flank can also cause subsidence of the summit crater. The crater on Mount Fuji, for example, is about 700 meters in diameter and 100 meters deep, with nearly vertical walls.

When a volcano erupts a very large volume of magma, the void created underground cannot support itself, and so part of the volcano subsides, or collapses, into the emptied space. The collapse creates a large circular basin with steep walls, sometimes several hundred meters high; many such basins are the remains of the summits of earlier volcanic cones. Such *collapse calderas* may form within a few hours or days, as at Krakatau and Pinatubo, and are generally associated with great eruptions (Figure 108).

The great Krakatau eruption in 1883 is a classic example of the scale and rapidity of caldera formation. Krakatau is an island volcano in the Straits of Sunda, between Java and Sumatra in Indonesia. These straits have been one of the principal shipping routes between Europe and Asia for centuries, and as a result, the volcanic activity of Krakatau has been well recorded in the logs of passing ships. In 1883, the summit of the island was lower than 1000 meters, and there had not been an eruption since 1680. On May 20, the eruption began with small explosions of ash from a low crater on the north end of the main island. The activity declined until late June, when passing ships reported two columns of steam. On August 26 the eruption increased dramatically. Sounds like thunder were heard at great distances, and by 2:00 P.M., a huge black cloud had climbed to 27 kilometers above the uninhabited island.

Aboard the *Charles Bal,* a British ship sailing north through the straits, the captain's log provides a vivid account of the climax of the great eruption. Here are a few excerpts from this extraordinary account:

> At 5 P.M. the roaring noise continued . . . darkness spread over the sky, and a hail of pumice stones began to fall upon us . . . The blinding fall of sand

107 Los Alamos, New Mexico, the birthplace of nuclear power, lies on the flank of a giant prehistoric volcano. The city sprawls across a canyon-cut plateau of thick pyroclastic flow deposits, erupted about 1 million years ago. Ejection of this huge volume of material caused the volcanic summit to collapse, forming Valle Grande, the 20-kilometer caldera covered by grasslands in the background. (Photograph by William Regan of the Los Alamos Scientific Laboratory.)

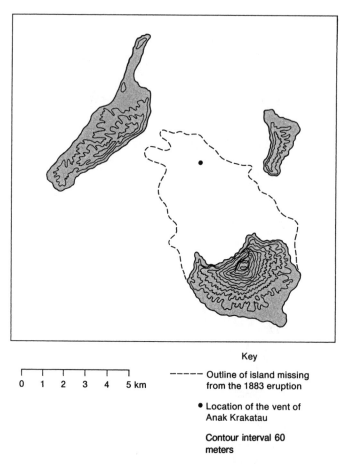

Key

0 1 2 3 4 5 km

– – – – – Outline of island missing
from the 1883 eruption

• Location of the vent of
Anak Krakatau

Contour interval 60
meters

108 Sketch map of the Krakatau Islands after the great 1883 eruption. The missing area is 23 square kilometers; where a volcanic peak 450 meters high once stood, the water is now 275 meters deep. Anak Krakatau—"child of Krakatau"—began erupting in 1927. An island appeared in 1930 and has grown by means of small explosive eruptions and lava flows to a height of 200 meters and a diameter of nearly 2 kilometers.

and stones, the intense blackness . . . broken only by the incessant glare of varied kinds of lightning . . . made our situation a truly awful one . . . At 11 P.M. . . . the island, eleven miles west-northwest, became more visible, chains of fire appearing to ascend and descend between the sky and it . . . The wind . . . was hot and choking, sulphurous, with a smell like burning cinders . . . From midnight to 4 A.M. . . . the roaring of Krakatoa . . . more explosive in sound, the sky one second intense blackness and the next a blaze of fire . . . At 11:15 there was a fearful explosion. . . . We saw a wave rush up on Button Island [50 kilometers northeast of the volcano] . . . The same waves seemed also to run right up on the Java shore. The sky rapidly covered in, by 11:30 A.M. we were enclosed in a darkness that might almost be felt. At the same time . . . a downpour of mud, sand, and I know not what . . . At noon the darkness was so intense that we had to grope about the decks, and although speaking to each other . . . yet could not see each other . . . By 2 P.M. we could see the yards aloft, and the fall of mud ceased. The ship . . . is as cemented; spars, sails, blocks and ropes in a terrible mess; but, thank God, nobody hurt or the ship damaged.

More than 36,000 people were drowned by the tsunamis from Krakatau on the low-lying coasts of Java and Sumatra. These giant waves were apparently generated by some combination of the sudden collapse of the sea bottom beneath the volcano and the great displacements of seawater by the huge pyroclastic flows disgorged by the eruption. Ships at sea near the volcano like the *Charles Bal* were not swamped by the great waves, but slowly rode up and down on the huge swells. As these waves reached shallow water, however, they began to curl and break, forming towering surf waves that swept inland, destroying everything in their reach.

The 10 A.M. explosion [11:15 A.M. on the *Charles Bal*'s clock?] was the largest natural explosion ever recorded. The ash cloud rose to 60 kilometers above Krakatau, and the detonation was heard as far away as Australia. Barographs around the world recorded the atmospheric pressure wave. Scientists began to recognize that we live on a small world subject to great natural events. In less than a day, Krakatau disgorged 18 cubic kilometers of rock and formed a circular caldera 6 kilometers in diameter and 1 kilometer deep. In that short time span, it erupted more volcanic rock than is formed along all the oceanic rifts in one year. Even though the Earth evolves slowly, much of the change is the sum of many brief, catastrophic moments.

Even greater caldera-collapse explosions have occurred in the past 200 years. In 1815 the eruption of Tambora Volcano, also in Indonesia, disgorged more than 100 cubic kilometers of volcanic material. In 1912 Katmai Volcano in Alaska erupted 20 cubic kilometers of ash and 10 cubic kilometers of pyroclastic flow deposits in two days, creating a caldera 5 kilometers wide. Going back into prehistoric time, Crater Lake in Oregon is another large collapse caldera. It formed during a great eruption that occurred 7700 years ago, as dated by the carbon 14 in charcoal from trees that were incinerated beneath the surrounding pyroclastic flows (Figure 109). Similarly, the 30-by-10-kilometer Lake Toba fills a giant caldera in Sumatra that collapsed about 75,000 years ago. Long Valley Caldera, near Mammoth Mountain, California, on the eastern side of the Sierra Nevada, extends 30 kilometers east–west and nearly 20 kilo-

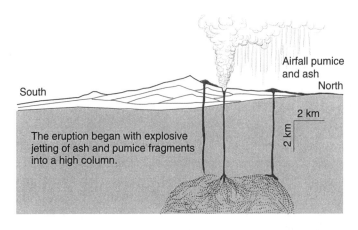

South   Airfall pumice
                                                               and ash
                                                                       North

                                                               2 km

The eruption began with explosive
jetting of ash and pumice fragments
into a high column.

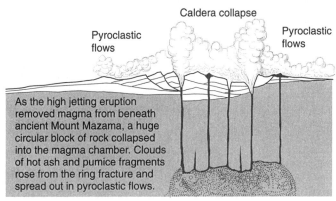

Caldera collapse

Pyroclastic   Pyroclastic
flows   flows

As the high jetting eruption
removed magma from beneath
ancient Mount Mazama, a huge
circular block of rock collapsed
into the magma chamber. Clouds
of hot ash and pumice fragments
rose from the ring fracture and
spread out in pyroclastic flows.

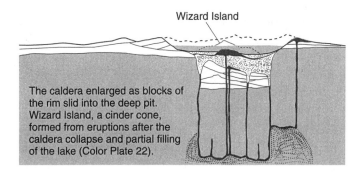

Wizard Island

The caldera enlarged as blocks of
the rim slid into the deep pit.
Wizard Island, a cinder cone,
formed from eruptions after the
caldera collapse and partial filling
of the lake (Color Plate 22).

109 Schematic cross sections showing the great eruption and caldera collapse that formed Crater Lake, Oregon, about 7700 years ago. (Drawings by Richard Hazlett, after Charles Bacon, U.S. Geological Survey.)

meters north–south. It apparently resulted from the collapse associated with the 700 cubic kilometers of pyroclastic flows that erupted violently some 700,000 years ago and deposited the apron of ignimbrites that surround the caldera.

Calderas are prominent features of several Hawaiian volcanoes. The summit of the major shield volcano Mauna Loa is the rim of a large elliptical caldera 3 by 5 kilometers in diameter, with vertical walls as high as 200 meters (Figure 110). This caldera formed about 1200 years ago,

110 Circular craters and a large 3-by-5-kilometer caldera indent the snow-covered summit of Mauna Loa Volcano in Hawaii. These collapse features, with cliffs up to 200 meters high, lie at an elevation of about 4000 meters, at the top of a huge, gently sloping shield volcano. Historic eruptions have been filling the craters and caldera with numerous flows. The view is to the north. (Photograph by the U.S. Army Air Force in 1939; courtesy of the National Archives.)

and the absence of explosive volcanic debris around its rim indicates that a major removal of magma from the summit chamber, probably by large flank eruptions of lava, caused the caldera to collapse. Although there were no eyewitnesses, it is inferred that the sinking of the Mauna Loa Caldera was incremental, and occurred slowly, in contrast to the explosive formation of many collapse calderas.

Craters and calderas are difficult to tell apart, and in fact, Kilauea Caldera is often called Kilauea Crater, even on official maps. A practical, though arbitrary, distinction can be made on the basis of size: craters are smaller than 1 kilometer in diameter, and calderas are larger than 1 kilometer in diameter. Another distinction is that calderas are generally formed by collapse (see Figure 109), while craters may be formed by either collapse or explosion.

Craters are sometimes hidden beneath domes of viscous lava; Lassen Peak in California is an example. Rows of small craters along a fissure erupting lava can be buried by the thick flows that issue from them. The lack of cones and craters on extensive lava plateaus such as the Columbia River lava fields in Washington and Oregon makes it difficult to locate the actual vents from which the lavas erupted.

Although volcanic mountains evolve through time, their mature and old-age forms are far from simple. A cinder cone may be the early stage of a large stratovolcano. Caldera collapse may then swallow up the summit of the stratovolcano, and new lava domes may subsequently appear over vents along the rim of the caldera. One way out of this classification dilemma is to call the result a *complex volcano* (Figure 111). Still, with careful geologic mapping, it is often possible to distinguish the component parts of a complex volcano and the order in which they formed. Such a history of eruptive habits and their sequence can be very useful in attempting to forecast the future hazards posed by a particular volcano.

Volcanoes can be erupting, dormant, or extinct; their life spans are extremely variable. The great volcanic complex near Los Alamos, New Mexico, has been erupting intermittently for 15 million years. Most individual volcanoes in Iceland erupt only once, suggesting that Surtsey's life span was only three and a half years. The average life span for a recurrently erupting volcano is roughly 100,000 to 1 million years. In the end, erosion takes over, and the stumps of vanquished volcanoes join the other passive mountains of the world.

The time of a volcano's death can be estimated by the degree of erosion, but such estimates must be made with caution. Erosion itself is extremely variable: fast in humid climates, slower in cold, dry regions, and fastest of all in cold, wet (glaciated) regions. The island of Hawaii has

111 An aerial view of Tongariro, Ngauruhoe, and Ruapehu volcanoes in New Zealand. The crater of Tongariro, in the foreground, is 1.3 kilometers wide and is transitional between a large crater and a small caldera. It is filled by ponded prehistoric flows cut by a small collapse crater. The cone of Ngauruhoe, in the middle ground, is typical of an active stratovolcano with a summit crater. Ruapehu, in the background, is also an active stratovolcano, but it contains a crater lake whose eruptions have prematurely eroded the volcano's flanks. (Photograph by S. N. Beatus, New Zealand Geological Survey.)

large variations in climate over short distances. The low slopes facing the trade winds are hot and humid and receive more than 5 meters of rainfall per year. The summit of Mauna Loa is cold and relatively dry, and on the lee shore, the climate is that of a desert. Lava flows only 50 years old have been reclaimed to soil and jungle in the hot, humid areas, but high on Mauna Loa, flows that are a thousand years old look freshly erupted. In the high deserts of Mexico, cinder cones remain almost untouched by erosion for thousands of years.

But climates change, and the tooth of time finally wears down all volcanoes. Since its formation a million years ago, half the Los Alamos plateau has been incised by deep canyons. Islands like Hawaii erode away in 5 million to 10 million years and then slowly subside beneath a growing coral atoll cap.

We have records of volcanic activity for 2000 years in Europe, 1000 years in Iceland, and only 200 years in the Cascade Range of the northwestern United States. Comparing these time spans with the million-year life span of volcanoes and the additional million years it takes to erode away the dead cones and craters, it is folly to declare a volcano extinct just because it has had no eruptions in recorded history. Any volcanic peak that shows little of the ravages of time, such as Mount Rainier, Mount Hood, or Mount Shasta, is only sleeping.

Links

Earth Observatory, NASA
http://earthobservatory.nasa.gov/NaturalHazards/natural_hazards_v2.php3?topic=volcano

Global Volcanism Program
http://www.volcano.si.edu/gvp/

Radar Images of Volcanoes, NASA
http://www.jpl.nasa.gov/radar/sircxsar/volcanoes.html

Stratovolcanoes, NOAA
http://www.ngdc.noaa.gov/seg/hazard/stratoguide/strato_home.html

Volcanic and Geologic Terms
http://volcano.und.nodak.edu/vwdocs/glossary.html

Volcano Data, NOAA
http://www.ngdc.noaa.gov/seg/hazard/volcano.shtml

Volcano Landforms
http://www.geology.sdsu.edu/how_volcanoes_work/Home.html

Volcano Slide Set, NOAA
http://www.ngdc.noaa.gov/seg/hazard/slideset/volcanoes/

Volcano World
http://volcano.und.edu

Volcanoes, Annenberg Channel
http://www.learner.org/exhibits/volcanoes/entry.html

Volcanoes.com
http://www.volcanoes.com/

Volcanology, NASA
http://eos.higp.hawaii.edu/index.html

Sources

Bullard, Fred. *Volcanoes of the Earth*. Austin: University of Texas Press, 1984.

Decker, Robert, and Barbara Decker. *Volcanoes in America's National Parks*. Hong Kong: Odyssey Guides, 2001.

Krafft, Maurice. *Volcanoes: Fire from the Earth*. New York: Abrams, 1993.

Scarth, Alwyn. *Volcanoes*. College Station: Texas A&M University Press, 1994.

Schmincke, Hans-Ulrich. *Volcanism*. Berlin: Springer, 2004.

Sigurdsson, Haraldur, ed. *Encyclopedia of Volcanoes*. San Diego, CA: Academic Press, 2000.

Winchester, Simon. *Krakatoa: The Day the World Exploded, August 27, 1883*. New York: Harper Collins, 2003.

Wood, Charles, and Jurgen Kienle, eds. *Volcanoes of North America*. Cambridge: Cambridge University Press, 1990.

CHAPTER **12**

Yellowstone

112 Old Faithful Geyser, 1871. This is one of the first classic photographs of Yellowstone taken by William Henry Jackson, the pioneering photographer who accompanied the Ferdinand Hayden expedition. (J. Paul Getty Museum, William Henry Jackson, "old Faithful," 1870, Albumen silver, 20 3/16" X 16 3/4".)

Yellowstone National Park was established in 1872, largely as a result of the vivid expedition report of Ferdinand Hayden and the sensational photographs of geysers and hot springs by pioneer photographer William Henry Jackson that he brought back to Washington. It is America's first national park, and it started the worldwide trend of preserving great scenic areas for future generations. Hayden recognized at once the volcanic nature of the area, but it was not until geologic mapping and radiometric dating of volcanic rocks in the second half of the twentieth century established the scale of Yellowstone's volcanic system that its enormous size was fully appreciated. It is now known as North America's largest volcanic system.

It was beaver pelts that lured the mountain men to this high plateau, and it was their tales of strange boiling springs and spouts of steam and hot water that lured Hayden to undertake his expedition. Yellowstone has the best and most extensive assortment of hydrothermal features in the world; Kamchatka, New Zealand, and Iceland also have beautiful hot springs and geysers, but Yellowstone is the winner, both in number and diversity. The word *geyser* derives from the Icelandic word *geysir*, which means "to gush." Geysers are rare; there are fewer than a thousand in the world. About 500 of them are in Yellowstone National Park, another 200 in Kamchatka, 40 in New Zealand, and 16 in Iceland. For each geyser there are many more hot springs.

The key to geyser operation is that the boiling point of water increases with pressure. In a narrow conduit of water beneath a hot spring, the surface boiling point of the water is controlled by altitude; for the high plateau of Yellowstone, it is about 92°C. Below the surface, the water pressure in the conduit increases from the weight of the water, and the boiling point increases rapidly to about 135°C at a depth of 30 meters. If the water at 30 meters depth continues to be heated until it reaches its boiling point, a bubble of steam forms and pushes some water out the top of the conduit. That starts a runaway reaction: as the water column in the conduit is lowered, the pressure at depth decreases, more steam is generated, more water gushes from the conduit at the surface, the pressure drops even more, and the runaway action of the geyser continues until the conduit is temporarily emptied.

Old Faithful Geyser is the icon of Yellowstone National Park (Figure 112), and is seen by about 3 million visitors every year. It erupts more frequently than any of the other big geysers, and its average repose time between displays varies from 65 to 92 minutes. Eruptions of 32,000 kilograms of steam and hot water, on average, last about one and one-half to five minutes. Longer eruptions are followed by longer repose intervals, which makes sense, since apparently it takes longer to reheat the additional water in the conduit. Rangers use these data to predict the next eruption time. The low hill at the site of Old Faithful is built up of siliceous sinter, a grayish-white mineral that precipitates from the erupted hot water as it cools. The size of the hill suggests that Old Faithful has been erupting for hundreds of thousands of years, not just since it was named by the Washburn expedition to Yellowstone in 1870.

The vent of the geyser is remarkably small, about 150 by 60 centimeters, and it narrows to a crack about 10 centimeters wide 6 meters down. This narrow point forms a nozzle that sprays the column of hot water and steam to heights of 30 to 55 meters. But these dry data do not do justice to the beauty and dynamic poetry of Old Faithful. If you have never seen it in action, plan a visit. Next best is a visit to the National Park Service Web site camera that presents live images of its eruptions.

Grand Prismatic Spring, a boardwalk attraction 15 kilometers north of Old Faithful, is one of the most dramatic and colorful of the thousands of hot springs at Yellowstone. A huge hot spring with a temperature of nearly 85°C, it flows an estimated 2100 liters per minute. Rings of colorful thermophile (heat-loving) cyanobacteria surround Grand Prismatic Spring in the shallow overflow zone at its edge. Different species of cyanobacteria, which prefer particular water chemistry and temperatures, form colored mats that provide a rough thermometer of the hot spring: yellow at 70°C; orange at 63°C; brown at about 55°C. The deep water inside the spring appears blue, fading to green in shallower water; "prismatic" is an excellent description of this magnificent hot spring.

It is cooling magma, emplaced a few kilometers underground, that powers these spectacular hydrothermal features. The very magnitude of the heat released at Yellowstone—30 to 40 times the average worldwide underground heat flow—is a major clue to the great size of this volcanic system. And yet, the youngest-dated lavas erupted at Yellowstone are

70,000 years old. The apparent explanation of this paradox is that Yellowstone's volcanic eruptions are huge, but not frequent, and that many shallow intrusions of magma help to maintain the system.

There were three gigantic prehistoric eruptions at Yellowstone. Enormous explosions of fragmented volcanic rock disgorged pyroclastic flows and ash columns and formed collapse calderas of fantastic scale. Twenty-five hundred cubic kilometers of volcanic explosion debris spewed from Yellowstone 2.1 million years ago, creating a caldera 90 kilometers long, 50 kilometers wide, and hundreds of meters deep. Thirteen thousand years ago, another great explosion from outside the present southwestern boundaries of the park erupted another 280 cubic kilometers of rhyolitic pumice and ash. The caldera it formed was more circular, about 16 kilometers in diameter. The youngest caldera, the present Yellowstone Caldera (Figure 113), 85 kilometers long and 45 kilometers wide, collapsed 640,000 years ago in still another explosion of 1000 cubic kilometers of rhyolitic volcanic fragments. The sum of these three great explosions, more than 3700 cubic kilometers of pyroclastic flows and volcanic ash, is more than 200 times the volume of the 1883 Krakatau eruption.

Ashfall from these huge eruptions was also prodigious, covering much of what is now the western United States from California nearly to the Mississippi River. If it were all swept together it would cover the present state of Wyoming with a layer 15 meters thick. The pumice blanket near the explosion vents was so hot that it partially melted, forming hard volcanic rock called *welded tuff*. It was mapping the extent and thickness of these deposits, largely the work of geologist Robert Christiansen of the U.S. Geological Survey (USGS), that revealed the true scale of the Yellowstone volcanic system.

From 160,000 to 70,000 years ago, after the present caldera was formed, 20 thick rhyolitic lava flows filled much of the depression (Figure 114). Many basaltic lava flows also occurred in the region 640,000–110,000 years ago. Perhaps most remarkable is that no volcanic deposits younger than 70,000 years have so far been discovered.

The geologic history of the Yellowstone system goes back much more than 2 million years. Seventeen million years ago, the Yellowstone hot spot was gestating and erupting a series of rhyolitic explosions and calderas in areas that are now located from northern Nevada through southern Idaho (Figure 115). As the North America Plate moves southwestward, the track of the relatively fixed location of the hot spot appears to move northeastward at a rate of about 3.5 centimeters per year. The ancient caldera centers are mainly buried by thick flows of basalt that followed the uplift and subsidence of the land surface over the hot spot. Bob Smith

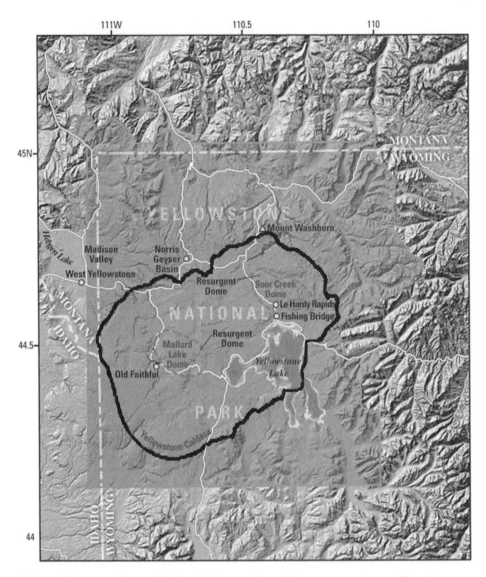

113 Yellowstone National Park, with the outline of the youngest giant caldera that erupted and collapsed 640,000 years ago. (U.S. Geological Survey.)

and Lee Siegel in their book *Windows into the Earth* make a wonderful analogy to describe the track of the hot spot: When the hot spot passed under an area, it created a 300-mile-wide, 1700-foot-tall bulge in the Earth's crust, like a rug being dragged over an object. So the old calderas—essentially "Old Yellowstones"—now hidden beneath the Snake River Plain once were at higher elevations. However, in the wake

114 Northwestern rim of Yellowstone Caldera north of Madison Junction. Much of the caldera floor (foreground) has been covered by younger rhyolitic flows. (Courtesy of Robert B. Smith, University of Utah.)

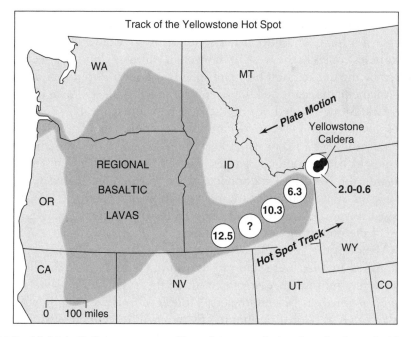

115 Multiple Yellowstone-type calderas form a track that dates back nearly 17 million years. This 600-kilometer-long "wake" of the Yellowstone hot spot begins near the present junction of Nevada, Oregon, and Idaho and continues east and northeast toward Yellowstone National Park. (Data from U.S. Geological Survey.)

of the hot spot, the plain sank as much as 2000 feet and was flooded by a series of basaltic lava flows, which covered the old calderas. What was left was a vast valley—the eastern Snake River Plain.

The origin of hot spots is controversial (see Chapter 8). Are they deep-rooted or shallow-rooted? Perhaps some are deep-seated, others shallow? Their origin and longevity are the topic of one of the liveliest geologic debates of our time. One thing is certain: they generate volcanoes, and if a tectonic plate moves over one of them, a long line of volcanic centers results. The Hawaiian hot spot has progressively spawned volcanoes for over 70 million years, and the line of oceanic basaltic islands it has produced—younger to the southeast—extends for 6000 kilometers. The line of ancient volcanic centers extending from the Yellowstone hot spot is 600 kilometers long and at least 16 million years old. One vast difference between the Hawaiian hot spot and the Yellowstone hot spot is their style of volcanism: effusive basalt in Hawaii; rhyolitic explosions at Yellowstone. This difference probably results from the fact that the Hawaiian volcanoes lie over thin oceanic crust, in contrast to the Yellowstone volcanoes, which lie over thick continental crust.

In the summer of 1969 a magnitude 7.5 earthquake violently shook the Yellowstone region. No one in the park was killed, but the shaking triggered a huge landslide in the Madison River Canyon near Hebgen Lake, a few miles to the west. Twenty-eight campers along the river were killed by the quake, most by a landslide that buried 19 of them beneath an avalanche that dammed the river. The shaking affected the geysers in the park. Many erupted during the next few hours and days; some new ones appeared, and even Old Faithful's schedule was perturbed. As often happens after a major natural catastrophe, new studies were initiated to study its background and causes. The University of Utah, in cooperation with the National Park Service, began research on the seismicity of Yellowstone by installing many new seismometers, and the USGS increased its research on the hydrothermal features and its geologic mapping of the prehistoric volcanic features.

As part of this research, new level surveys of a series of benchmarks established in 1923 were done in the summers of 1975–1977. The results were astonishing. A large area inside the Yellowstone Caldera had bulged as much as 72 centimeters. More level surveys using new techniques, such as Global Positioning System (GPS) technology and satellite radar interferometry (InSAR; see Chapter 18), followed, and the complex ups and downs of the caldera became apparent. Until 1985 the caldera floor continued to rise a few centimeters per year. Then, from

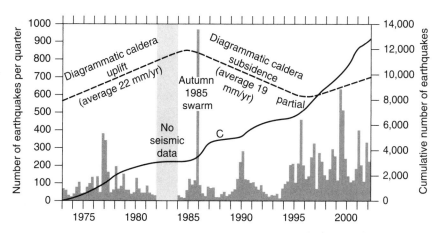

116 Uplift and subsidence of part of the floor of Yellowstone Caldera from 1973 to 2002, and the number of small, local earthquakes over the same time period. (Data from U.S. Geological Survey.)

1985 to 1995, it subsided at about the same rate, after which uplift resumed (Figure 116). Multiple ideas were put forth to explain this complex behavior: first, that additions and withdrawals of magma in the subsurface were causing the ups and downs; second, that volcanic gas escaping from deeper magma was causing pressure changes as it was accumulated and released from beneath the deforming areas; and third, that the hot water trapped beneath the region was increasing in temperature, pushing upward until it was released by leakage from the system. Whatever the cause or causes, it is clear from the ongoing seismicity and deformation that Yellowstone is an active volcanic system despite its apparent 70,000-year hiatus of eruptions (Figure 117).

Is there danger from volcanic hazards to the millions of visitors to Yellowstone National Park? Not much. The chance of stepping into scalding water by not staying on the boardwalks is much, much higher than that of being blasted by some future caldera eruption. The human life span is so short compared with the pace of geologic time that the danger is minimal, but it is not zero. In part to monitor the potential danger, but even more to try to understand this giant volcanic system, the USGS, the University of Utah, and Yellowstone National Park in 2001 established the Yellowstone Volcano Observatory. Its Web site provides up-to-the-minute observations on seismic activity beneath the park and longer-term data and interpretations on deformation and hydrothermal features.

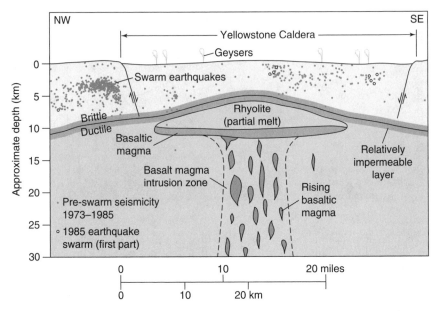

117   Schematic cross section beneath Yellowstone Caldera. Rising blobs of basaltic magma melt the continental crust, creating a magma chamber of partly molten rhyolite. (Data from U.S. Geological Survey, 2004.)

Besides being America's largest volcanic system, Yellowstone is an entire ecosystem large enough to maintain itself in its nearly pristine form. Wildlife, forests, clear streams, waterfalls, and mountain air are not greatly different from when Hayden and Jackson first gazed on this unique area.

Links

Geyser Observation and Study Association
http://www.geyserstudy.org/

Old Faithful Cam
http://www.nps.gov/yell/oldfaithfulcam.htm

U.S. Geological Survey, Volcanoes
http://volcanoes.usgs.gov/

Tracking Changes in Yellowstone's Restless Volcanic System
http://pubs.usgs.gov/fs/fs100-03/

The Total Yellowstone Page
http://www.yellowstone-natl-park.com/

Yellowstone Database
http://www.wsulibs.wsu.edu/yellowstone/

Yellowstone Earthquakes
http://www.seis.utah.edu/HTML/YPSeismicityMaps.html

Yellowstone National Park.Com
http://www.yellowstonenationalpark.com/index.htm

Yellowstone National Park, NPS
http://www.nps.gov/yell

Yellowstone Volcanic Observatory
http://volcanoes.usgs.gov/yvo/

Sources

Christiansen, R. L. *The Quaternary and Pliocene Yellowstone Plateau Volcanic Field of Wyoming, Idaho, and Montana.* U.S. Geological Survey Professional Paper 729-G, 2001. 145 pp., 3 plates.

Christiansen, R. L., and R. A. Hutchinson. "Rhyolite–Basalt Volcanism of the Yellowstone Plateau and Hydrothermal Activity of Yellowstone National Park, Wyoming." In *Centennial Field Guide*, vol. 2, pp. 165–172. Boulder, CO: Geological Society of America, 1987.

Dzurisin, D., C. J. Wicks, Jr., and W. Thatcher. "Renewed Uplift at the Yellowstone Caldera Measured by Leveling Surveys and Satellite Radar Interferometry." *Bulletin of Volcanology* 61 (1999): 349–355.

Fournier, R. O. "Geochemistry and Dynamics of the Yellowstone National Park Hydrothermal System." *Annual Review of Earth and Planetary Sciences* 17 (1989): 13–53.

Fournier, R. O. "Old Faithful: A Physical Model." *Science* 163 (1969): 304–305.

Humphreys, G., D. Schutz, K. Dueker, and R. B. Smith. "Plume or No Plume at Yellowstone?" *GSA Today* 10, no. 12 (2000): 1–7.

Husen, S., R. B. Smith, and G. P. Waite. "Evidence for Gas and Magmatic Sources beneath the Yellowstone Volcanic Field from Seismic Tomographic Imaging." *Journal of Volcanology and Geothermal Research* 131 (2004): 397–410.

Pierce, K. L., and L. A. Morgan. 1992. "The Track of the Yellowstone Hot Spot: Volcanism, Faulting, and Uplift." In P. K. Link, M. A. Kuntz, and L. B. Platt, eds., *Regional Geology of Eastern Idaho and Western Wyoming*, pp. 1–53. Boulder, CO: Geological Society of America, 1992.

Smith, R. B., and L. W. Braile. "The Yellowstone Hotspot." *Journal of Volcanology and Geothermal Research* 61 (1994): 121–188.

Smith, R. B., and L. Siegel. *Windows into the Earth: The Geologic Story of Yellowstone and Grand Teton National Park.* New York: Oxford University Press, 2000.

Waite, G. R., and R. B. Smith. "Seismic Evidence for Fluid Migration Accompanying Subsidence of the Yellowstone Caldera." *Journal of Geophysical Research* 107, no. B9 (2002), 2177, 10.1029/2001JB000586.

CHAPTER 13

Roots of Volcanoes

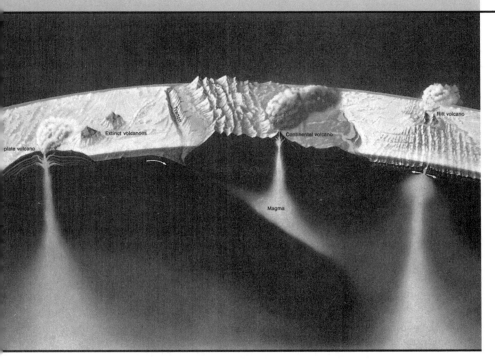

118 A schematic diagram of the three types of volcanoes related to plate tectonic processes: a rift volcano formed by separating plates, like the submarine volcanoes on the Mid-Atlantic Ridge and Icelandic volcanoes (right); a volcano formed by converging plates at subduction zones, like those of Japan and the Andes of South America ("continental volcano," center); and a hot-spot volcano, like those of Hawaii and the Hawaiian Ridge ("mid-plate volcano," left). (From *Powers of Nature*. Copyright National Geographic Society. All rights reserved.)

W hen John Wesley Powell was trying to establish the U.S. Geological Survey, a senator from Nebraska challenged him, saying: "You can't see any farther into the Earth than any other man." Even today, geologists can hardly deny the truth of that statement, but we do have a great deal of indirect data on the inside of the Earth that helps guide our speculations.

For example, we can study the seismic waves from large earthquakes because they travel through the Earth to seismographs around the entire globe. Earthquake waves change in character as they travel away from their source, but regardless of the location of the earthquake, seismographic recordings made at equal distances from the source are nearly the same. This result can best be explained by an interior structure of the Earth that is symmetrical around the center (Figure 119); thus, if the

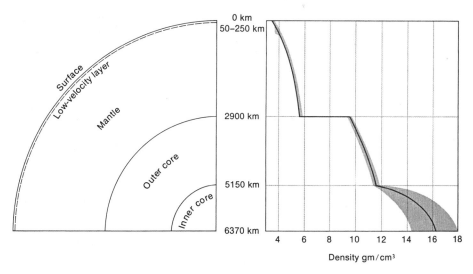

119 The estimated density of the interior of the Earth. The solid line in the graph on the right shows the most probable value of density at each depth, and the shaded region outlines the range of uncertainty. (Data from K. E. Bullen, "The Interior of the Earth." Copyright © 1955 by Scientific American, Inc. All rights reserved.)

Earth's interior is composed of various materials of different densities, they must be arranged in layers or spherical shells like an onion.

We also have clues to the weight of the Earth's interior. We know the average density of the whole Earth from its size and from the magnitude of the force of gravity at its surface: it amounts to 5.5 grams per cubic centimeter, which is twice the density of surface rocks. Therefore, some parts of the Earth's interior must be very dense material to bring the average up to 5.5.

Reasonable speculations about the roots of volcanoes must be consistent with our knowledge of the physics and chemistry of the whole Earth as well as our observations of the surface geology. One fact is quite clear: the surface geology indicates that our Earth is very complex and heterogeneous. The physical and chemical data on the Earth's interior yield average values rather than details, however, so most models of the Earth's interior show simple shells of rock or molten metal. As a first approximation these cartoons are probably accurate, but the details are missing. Filling in this picture is the most fascinating challenge in geoscience research today.

These details are not trivial. Diamond and graphite are the same chemical, but wars have been fought over their difference. Synthetic diamonds can be made from graphite in complex high-pressure, high-temperature furnaces reaching pressures of 100,000 atmospheres (103,320 kilograms per square centimeter) at more than 1000°C. Natural diamonds occur in volcanic pipes that originated at depths as much as 200 kilometers beneath the surface; thus the diamonds, their inclusions, and the pieces of wall rock erupted with them provide a direct sample of the composition, pressure, and temperature at depths in the Earth far below the reach of drills (Figure 120).

Uplift and erosion have exposed the shallow roots of ancient volcanoes that were once as much as 10 kilometers below the surface. Postmortems on the dikes and pipes of chilled magma that remain show that in many of these volcanoes, the surface vents were connected to larger storage chambers of molten rock at depths of 2 to 10 kilometers beneath the surface (Figure 121). These shallow magma chambers were complex bodies of intersecting dikes and layers of molten rock or, sometimes, more cylindrical masses called *stocks* or *plutons*. Their composition, the texture of their crystals, and the host rocks at their edges yield important clues to how the magma chambers evolved and how and when they cooled into crystalline rocks. In general, the composition of shallow magma chambers is the same as that of surface volcanic products, although their texture is more coarsely crystalline because they cooled more

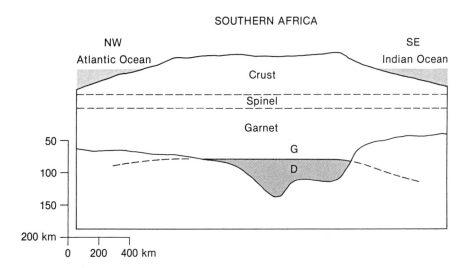

120 A schematic cross section through southern Africa, based on inclusions brought up in diamond pipes and other volcanic pipes in which diamonds are absent. Spinel and garnet are minerals characteristic of the depth ranges indicated. The boundary labeled G/D is the transition between graphite and diamonds. (Modified from Francis R. Boyd, Carnegie Institution, *Year Book 1986*, p. 98.)

slowly. For example, granite is the root rock in stocks beneath volcanoes that once erupted rhyolitic lava flows or huge volumes of rhyolitic pyroclastic flows (see Table 2).

These shallow magma chambers cannot be the deep roots of volcanoes, however, because they occur at depths too shallow for rocks to be melted by the Earth's heat. Uplift and erosion have not exposed the deep roots of volcanoes; they are still hidden from direct observation, so our information about them is more tenuous. The source of the heat that melts rock was once a subject of lively debate. The leading contenders were the Earth's original heat from the time of formation; heat from the breakdown of the radioactive isotopes of uranium, thorium, and potassium; heat from the gravitational energy of the redistribution of heavy elements toward the center of the Earth; and heat from the tidal friction produced by the slowing of the Earth's rotation. Most geologists now agree that the decay of radioactive isotopes is the principal heat source.

The accretion of the planets at the time the solar system was formed generated immense amounts of heat from collision and gravitational collapse. Whether or not the planets actually melted depended on the speed of their accretion. They would have melted more completely if they had formed rapidly, because the heat would have had less time to be lost to

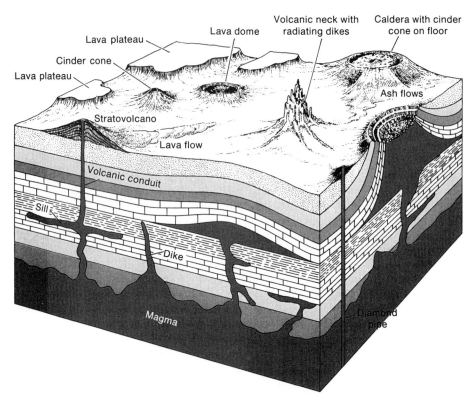

121   A schematic diagram of the surface forms and subsurface structure of various volcanic features. (Adapted from R. G. Schmidt and H. R. Shaw, *Atlas of Volcanic Phenomena*, U.S. Geological Survey)

space. Even if the original Earth were molten, however, the heat of its formation alone could not account for the estimated heat lost during its 4.5-billion-year life span.

Moreover, even if the Earth had formed cold, the radioactive decay of unstable atoms would have started a warming trend—warming rapidly at first because of the many short-lived, highly radioactive elements and then more slowly from the longer-lived but less radioactive elements that remained, such as uranium 238, thorium 232, and potassium 40. The heat generated from the estimated remainder of these three isotopes in the Earth today can account for all of the world's currently escaping heat (Table 4).

As the primitive Earth was initially being heated by radioactive decay, melting and the gravitational segregation of iron to form the core would have greatly hastened the heating process, perhaps melting the en-

TABLE 4	Radioactive heat production from uranium, thorium, and potassium in common igneous rocks			
Rock	Amount of Radioactive Element in Rock (parts per million)			Amount of Heat Produced (joules/kilogram year)
	Uranium	Thorium	Potassium	
Granite	4	13	4	.03
Basalt	0.5	2	1.5	.005
Peridotite	0.02	0.06	0.02	.0001

tire Earth. Even so, this gravitational melting would have been a one-time process and its heat long gone unless sustained by other processes.

The heat produced by tidal friction gets its energy from the slowing down of the Earth's rotation. (Some studies of the growth bands of 400-million-year-old fossil corals indicate that the early Earth's rotation was faster and the days shorter.) Much of this energy is dissipated in the swirling tides of the Earth's oceans, but some geologists believe that the tidal warping of weak zones inside the solid Earth may be another continuing source of heat. As in many geologic debates, the problem is complicated by too many plausible explanations. Probably more than one of the various theoretical sources of energy are involved, and all of them are virtually inexhaustible in terms of human time spans.

Regardless of the exact source of the heat in the Earth, much is known about the variation in temperature near its surface. In deep mines and drill holes, the temperature increases with depth. The rate of this increase, called the *thermal gradient*, averages about 30°C per kilometer of depth. By measuring the insulating quality of rocks, which is quite high, the actual loss of heat from inside the Earth can be calculated. This quantity is about 5000 times less than the heat that reaches the Earth from the sun. Even so, it is an enormous amount of energy, much greater than the more spectacular heat loss from all the Earth's volcanoes.

At 30°C per kilometer, the 800°C–1200°C melting temperatures of rock should be reached at depths of 30 to 40 kilometers beneath the Earth's surface. There is good evidence, however, that the thermal gradient is not constant, but rather slows at greater depths, because of the way that the radioactive isotopes of uranium, thorium, and potassium are distributed in the Earth. These elements have an affinity for granitic rocks and tend to be concentrated in the Earth's crust. The production of heat from their radioactive decay therefore diminishes with depth, reducing the rate at which temperatures increase with depth. The best estimates of this effect indicate that temperatures inside the Earth reach 800°C–1200°C at depths of 60 to 100 kilometers beneath the surface. This is the same depth as the top of the seismic low-velocity layer (the asthenosphere) described in Chapter 1—an important piece of independent evidence supporting the temperature calculations. The roots of volcanoes must therefore extend down at least 60 to 100 kilometers beneath the surface to reach a major source of magma (Figure 122).

Seismic evidence indicates that the low-velocity layer is not a zone of completely liquid rock. Rather, the rock is only partially molten; perhaps a small percentage of the total material is liquid and is contained in a spongelike mass of weak but solid rock at a high temperature. How, then, does the molten magma separate and rise toward the surface?

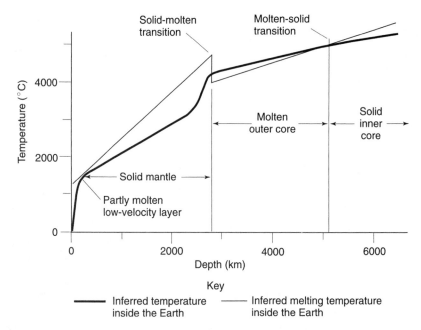

122 A graph of inferred temperature (black line) and the inferred melting temperature (gray line) of rock inside the Earth. There are four control points for this graph: (1) The thermal gradient near the Earth's surface is about 30°C per kilometer of depth. (2) At a depth of about 100 kilometers, the actual temperature and melting temperature curves must almost touch to account for the low strength and partial melting of the low-velocity layer. (3) For the lower mantle to be solid and the outer core molten, the actual temperature and melting temperature curves must cross. (4) For the outer core to be molten and the inner core to be solid, the actual temperature and melting temperature curves must again cross. (Data from Reinhard Boehler, 1996.)

The density of magma is less than that of the rock from which it melts. Under the influence of gravity, the lighter molten material tends to rise and the residual solid material to sink. Any fractures reaching deep into the Earth would hasten this upward movement of magma, which would tend to escape in the same manner as petroleum gushes from a wild well. The problem is the conduit. Do fractures reach down through the rigid crust into the melt zone, or do bodies of magma push and melt their way slowly upward as great rising blobs to feed the shallow magma chambers (Figure 123)?

Divergent plate margins have major fractures, and volcanoes along the mid-ocean ridges probably have conduits along these fractures between the plates. In these rift zones the thermal gradients are much higher than average, and seismic evidence indicates that the low-velocity layer comes up to within 10 or 20 kilometers of the surface.

At convergent plate margins, subduction volcanoes occur in belts parallel to the ocean trenches. The volcanic belt is often, but not always, offset 100 to 200 kilometers onto the overriding plate (see Figure 56). The generation of magma seems to be aided by water and carbon dioxide from the subducting slab where it reaches a depth of 100 to 150 kilometers. This variation in offset argues for magma's forcing its way upward rather than rising along some special fracture zone.

In Hawaii, the location of earthquakes below the shallow magma chamber of Kilauea indicates that there is a zone of fractures shaped like an inverted funnel about 5 to 50 kilometers in diameter and 10 to 50 kilometers deep, reaching down to the top of the low-velocity layer. The shape of this region of active cracking does not imply a major fracture zone parallel to the Hawaiian Island chain, but rather a local zone of failure closely associated with the ascending magma. Here it appears that the magma forces its own way up through many small and disconnected fractures to the shallow magma chamber beneath Kilauea. The rate of inflation of the summit area during the years from 1960 to 1986 indicated that magma was being fed continuously from depth into the magma chamber at 3 to 6 kilometers below the surface, at rates of about 100,000 to 500,000 cubic meters per day. From 1986 through 2003, Pu'u O'o has been erupting in a nearly steady-state manner, and the summit of Kilauea has shown only a few brief periods of inflation.

No one has ever seen the deep roots of volcanoes, but improved geophysical techniques can provide some startling indirect views. In Yellowstone National Park, small delays in the transit time of vibrations from distant earthquakes indicate that a large body of magma or very hot rock underlies much of the region (Figure 124). By recording these delays of only seconds at many seismograph locations, geophysicists can map the rough size and shape of the huge thermal anomaly.

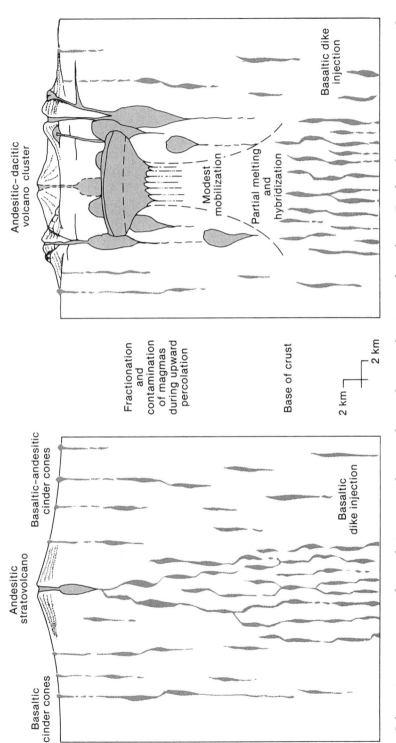

123 Schematic cross sections of an andesitic stratovolcano and an andesitic-to-dacitic complex volcano. Basalt is the primary magma rising from depth, but it becomes more silica-rich by the removal of early-formed crystals (fractionation) and by contamination of the basalt with crustal rocks of higher silica content. Andesite-to-dacitic magma chambers are shown by the dotted pattern. (From Wes Hildreth, *Journal of Geophysical Research* 86 [1981]: 10179.)

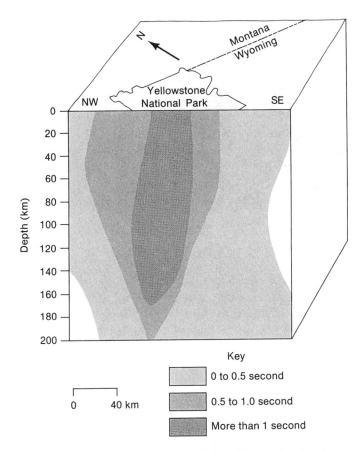

124   A schematic diagram of Yellowstone National Park, showing the seismic velocity anomaly at depth. Vibrations from distant earthquakes are delayed more than 1 second during their passage through the region shown. The carrot-shaped zone may include a huge volume of still-molten rock. (Data from H. M. Iyer, U.S. Geological Survey.)

Another seismic technique, a spin-off from prospecting for oil by generating sound waves at the Earth's surface and studying their reflections, extends the "X-ray vision" of geophysicists to a depth of 50 kilometers. Analysis of the deep seismic cross sections obtained by this technique reveals a great deal of complexity in the rock structures at depth. In the Rio Grande Rift section of New Mexico, a major sound-reflecting zone at about 20 kilometers beneath the surface is thought to be a layer of magma. An adaptation of this technique to seismic exploration beneath the seafloor reveals the top of a shallow magma chamber along the axis of the East Pacific Rise (Figure 125).

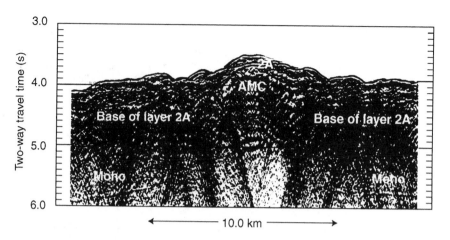

125 A seismic reflection profile of the crest of the East Pacific Rise near 14°14′ south. The reflection labeled 2A, at a depth of about 200 meters below the seafloor beneath the ridge axis, is interpreted to come from the base of the lava flows that overlie dikes. The strong reflection about 1 kilometer below the seafloor, labeled AMC (axial magma chamber), is interpreted to come from the top of a thin layer of molten rock. The base of the oceanic crust, labeled Moho, lies at a depth of about 7 to 8 kilometers below the seafloor. (From Robert Detrick et al., *Science* 259 [22 January 1993]: 501.)

A new branch of geophysics called *seismic tomography* uses the travel times of earthquake waves that have passed deep within the Earth's interior. By comparing the actual travel times with those for a uniformly layered model of the Earth and integrating the data from hundreds of earthquakes and thousands of seismograms, geophysicists are able to draw three-dimensional plots of some of the larger thermal and compositional anomalies in the major layers of the Earth's interior. As is usually true, the closer you look at a feature, the more complex it turns out to be. The once-simple circle that separated the Earth's core and mantle is now being redrawn into complex compositional and thermal boundary layers. For geologists and volcanologists who are getting their first geophysical look at magma and the deep roots of hot spots, these are exciting times.

Links

Adam Dziewonski's research projects
  http://www.seismology.harvard.edu/~dziewons/

Cooperative Institute for Deep Earth Research (CIDER)
  http://www.deep-earth.org/

Don Anderson
  http://www.gps.caltech.edu/~dla/

International Association of Seismology and Physics of the Earth's Interior (IASPEI)
http://www.iaspei.org/home.html

International Heat Flow Commission
http://www.geo.lsa.umich.edu/IHFC/

Mantle dynamics, etcetera
http://cfauvcs5.harvard.edu/lana/lana/index.htm

Origin of "Hotspot" Volcanism
http://www.mantleplumes.org/

Scripps mantle tomographic models
http://mahi.ucsd.edu/Gabi/3dmodels.html

Seismo surfing
http://www.geophys.washington.edu/seismosurfing.html

Sources

Ahrens, Thomas, ed. *Global Earth Physics*. AGU Reference Shelf, no. 1. Washington, DC: American Geophysical Union, 1995.

Anderson, D. L. "Top-Down Tectonics?" *Science* 293 (2001): 2016–2018.

Anderson, Don, T. Tanimoto, and Y. Zhang. "Plate Tectonics and Hotspots: The Third Dimension." *Science* 256 (1992): 1645–1651.

Boehler, R. "Melting Temperature of Earth's Mantle and Core: Earth's Thermal Structure." *Annual Review of Earth and Planetary Sciences* 24 (1996): 15–40.

de Silva, S. L. "Altiplano–Puna Volcanic Complex of the Central Andes." *Geology* 17 (1989): 1102–1106.

Dziewonski, A. M., A. M. Forte, W.-J. Su, and R. L. Woodward. "Seismic Tomography and Geodynamics." In K. Aki and R. Dmowska, eds., *Relating Geophysical Structures and Processes: The Jeffreys Volume*, 67–105. Geophysical Monograph 76. Washington, DC: American Geophysical Union, 1993.

Lee, William, ed. *International Handbook of Earthquake and Engineering Seismology*. Amsterdam: Elsevier, 2002.

Wyllie, Peter (chair), and the Committee on Status and Research Objectives in the Solid-Earth Sciences. *Solid-Earth Sciences and Society*. Washington, DC: National Academy Press, 1993.

CHAPTER **14**

Origin of the Sea and Air

126 Sea, steam, and flying rocks: a Hawaiian lava flow enters the ocean. (Photograph by G. A. Macdonald, U.S. Geological Survey.)

The smell of sulphur is strong, but not unpleasant for a sinner.
—Mark Twain (1866)

The brimstone smell of sulfur dioxide at active volcanoes is noxious, especially to people with respiratory problems, but usually not fatal. Another gas commonly exhaled at volcanic vents is more insidious, however. Carbon dioxide—a colorless, odorless gas—is heavier than air and can accumulate in concentrations that can suffocate animals and people with little or no warning. A tragic example occurred at Lake Nyos in Cameroon, West Africa, on August 21, 1986.

Lake Nyos fills a 1- to 2-kilometer-wide, 200-meter-deep crater on a dormant volcano that had not erupted in recorded history. Without warning, a rumbling burst of gas, mostly carbon dioxide, belched from the lake and poured into the valley below. A survivor who lived above the lake heard the rumbling, and in the darkening evening light saw a gas cloud that looked like a smoking river moving down into the valley. He felt a blast of wind from the lake and was gagged by its awful smell.

The dense cloud of gas—later estimated to be as much as 1 cubic kilometer in volume—flowed down the valley below the crater for 10 kilometers, killing 1700 people and 3000 head of cattle. Houses and farms were undamaged, but dead bodies were everywhere.

Scientists who investigated the disaster agree that carbon dioxide gas of volcanic origin was the killer, although they disagree about just what happened. Most of them think that for hundreds of years, magmatic gases had been seeping into the lake from below and, because of the high pressure of the overlying water, were dissolved and concentrated in the deep water near the lake bottom. This rare phenomenon can occur in a tropical lake because the warmer, less dense layer of water near the surface prevents the lake water from overturning, which would release most of the gases dissolved at depth. In temperate regions, lake water also forms a warmer, less dense layer on top of the cold bottom water during the summer, but cooling of the surface water in winter eliminates this layering, and wind and currents bring the deep water to the surface, so that any accumulated gas is released harmlessly. Once the deep water in a tropical crater lake has become saturated with carbon dioxide, the potential for its rapid effervescence—like the sudden opening of a champagne bottle—exists.

Although most scientists agree up to this point, they disagree on what triggered the sudden release of dissolved gas from Lake Nyos. Some believe that large amounts of cool rainwater entering the lake or a landslide on the steep wall of the lake basin could have triggered the release of gas. Once the effervescence was started, it would stir the lake and continue the rapid degassing. Another school of thought holds that a small volcanic eruption beneath the lake triggered the gas outburst, and a few investigators believe that all the carbon dioxide was of sudden eruptive origin. No new volcanic rock has been found around the lake, but evidence for this interpretation may lie hidden at the bottom of now-calm Lake Nyos. The recent experimental emplacement of a pipe from the surface down into the depths of the lake has produced a spectacular jet of CO_2-laden water. This observation supports the slow accumulation argument, and also indicates a way to prevent another gas-release disaster.

But catastrophic events like this one are rare. In a larger view, over geologic time, volcanic gases have helped make the Earth—with its abundant surface water—a special planet in our solar system (Figure 126). Of all the inner planets—Mercury, Venus, Earth, and Mars—only Earth has oceans. Venus has a massive, dense atmosphere of carbon dioxide and sulfuric acid that perpetually hides its rocky face in clouds, but its 500°C surface temperature is too hot to allow the condensation of an ocean.

What is the origin of the water and gases on Earth, and why are the inner planets so diverse in this respect? Many factors seem to be involved: the planet's distance from the sun, the composition of the nebular gases from which it condensed, the planet's mass, the presence or absence of an original atmosphere, the changing temperature of its interior, and the amount of water and other gases bound up with the rocks and metals that formed the planet. The distance from the sun strongly affects a planet's surface temperature, and the planet's mass controls its gravity. Light gases escape into space when the surface is warm and the gravity low. For these reasons, the moon's surface is a vacuum: it does not have enough mass to hold an atmosphere at its distance from the sun.

The question of how much of the Earth's atmosphere was originally present—that is, formed at the time the planet formed—is the basis of a controversial and lively debate that also involves the role of volcanoes. If the Earth formed slowly by accretion from a cosmic dust cloud, it may have remained cool enough for an original atmosphere to form. By contrast, if the accretion was rapid and hot, or if the surface temperatures were higher than at present because of either internal processes, such as radioactive heating, or external processes, such as excessive heat from the early ignition of the sun, any original atmosphere would have boiled away.

The "neon argument" supports the concept of the loss or lack of any original atmosphere. Neon is an inert gas; it does not combine with silicate minerals or metals to form solid compounds. Neon is also relatively abundant as a gas in the cosmos, and its density indicates that it is not subject to loss into space from the Earth's present atmosphere. An original Earth atmosphere derived from a cosmic dust cloud should have contained a modest amount of neon. But the Earth's present atmosphere has only a tiny fraction of neon, much less than would be expected in an original atmosphere. Its near absence indicates that the original neon was lost, along with all the other gases and water, from the hot surface of a primitive Earth. A later atmosphere and oceans evolving from volcanic gases escaping to the Earth's surface would not include much neon. Only gases such as water, carbon dioxide, and others that can be chemically bound into solid matter would be released by melting and eruption on the Earth's surface.

For these reasons, many geologists think that at least part of the Earth's oceans and atmosphere boiled out of the Earth's interior, particularly during the period of melting resulting from the conversion of gravitational energy as its dense core formed. Alternatively, collisions with the abundant icy comets that are thought to have populated the early solar system may have been the major contributor of water and carbon dioxide to the young Earth's surface.

Before the concept of plate tectonics was firmly established, it was thought that volcanic gases were the main source of the Earth's oceans and atmosphere, which were formed by a long evolutionary process. Now this thinking has shifted, and volcanoes are viewed as recycling the oceans and atmosphere instead of creating them. The general argument goes like this: Water, carbon dioxide (in calcium carbonate sediments), and other volatile compounds (that is, those that form gases at low temperatures and pressures) are carried down into the mantle on seafloor crust at subduction zones. After a few million years these volatiles are exhaled by eruption at subduction volcanoes or, after a few hundred million years, by eruption at rift volcanoes. Hot-spot volcanoes may also recycle some of the volatiles incorporated into the low-velocity layer at subduction zones, but the ratio of helium isotopes in hot-spot volcanic gases suggests that they have a deeper, more primitive source.

With this idea in mind, let's look at the actual data on gases escaping at active volcanoes. Molten basalt rising to the Earth's surface contains about 0.5 to 1 percent, by weight, dissolved gases. In sharp contrast, silica-rich magmas have a much more variable gas content and can contain as much as several percent gas, by weight. These gases are dissolved in the magma in the same way that carbon dioxide gas is dissolved in beer. They escape from solution by effervescence into gas bubbles and foam as the magma reaches the surface, just as beer foams when its containing pressure is suddenly released. The amount of gas dissolved in magma is the principal factor controlling the explosiveness of eruptions.

It is not easy to determine the composition of volcanic gases, since they escape during eruptions. Huge volumes of gases are involved in explosive eruptions, but it is nearly impossible to get close enough to sample them before they mix and react with unknown amounts of air (Figure 127). Indirect methods of sampling have proved more successful. Gases in deep submarine eruptions are frozen into the glassy rinds of pillow lava. These trapped gases have been analyzed by several investigators and provide some of the best estimates of the composition of volcanic gases currently available.

Sometimes crystals forming in a cooling magma chamber enclose a tiny pocket of melt. If these crystals are erupted, the melt pocket cools rapidly to an inclusion of glass in which the volcanic gases are trapped. An analysis of these inclusions by Fred Anderson at the University of Chicago revealed that magma from subduction volcanoes has a much higher gas content than does magma from submarine rift volcanoes.

Richard Stoiber of Dartmouth College pioneered still other techniques for analyzing volcanic gases. He used instruments called *absorption spectrometers* to measure the sulfur dioxide in volcanic fume clouds backlighted by the sky. Stoiber also analyzed gases adsorbed on volcanic ash particles collected as they fell from eruption clouds. His studies indicated that each active volcano generates about as much sulfur gas pollution as does a large coal-fired electric power station. Table 5 gives Stoiber's estimates of the annual emission of major gases from volcanoes on land.

A volcanic gas seldom makes headlines, but several newspaper articles have recently been published about the emission of carbon dioxide seeping up through the ground near Mammoth Mountain, a popular ski resort in eastern California. This seepage of gas was first suspected in 1990 when localities were found in which all the trees had been killed, and it was confirmed when a forest ranger almost died from suffocation inside a snow-survey cabin. Since 1980, the general area—a giant caldera that collapsed 700,000 years ago—has been undergoing earthquake swarms and uplift (Figure 128). The last eruption occurred here 500 years ago and was small compared with the giant eruption that created the caldera, but it definitely establishes that Long Valley Caldera is only dormant, not extinct. Geologists monitoring the earthquakes, uplift, and now emission of carbon dioxide are not forecasting an imminent eruption,

127 Volcanic gases escaping from this crack in a cinder cone formed during an eruption of Kilauea Volcano in Hawaii are depositing an encrustation of minerals. Wayne Ault of the Hawaiian Volcano Observatory measures the temperature of the vent and collects gas for chemical analysis. (Photograph by the U.S. Geological Survey.)

TABLE 5   Estimated gas emissions
            from world's volcanoes
            on land

Gas	10^{12} Grams/Year
Water	713
Carbon dioxide	65
Sulfur dioxide	19
Hydrogen chloride	3
Hydrogen fluoride	0.1

Source: Adapted from Richard Stoiber, "Volcanic Gases from Subaerial Volcanoes on Earth," in Thomas Ahrens, ed., *Global Earth Physics*, AGU Reference Shelf, no. 1, Washington, DC: American Geophysical Union, 1995.

but they do think that dikes of magma have been injected into the rocks beneath Mammoth Mountain. Carbon dioxide is apparently escaping from this magma and seeping to the surface. Here is a classic standoff between the scientists investigating a restless caldera and the residents of Mammoth Lakes, a town supported by a ski resort and summer recreation. The investigators want to understand how volcanoes work and how to reduce volcanic risk; many residents wish they would do it somewhere else.

Better measurements and estimates of the amounts of volcanic gases on a worldwide basis are still being made. Some recent results from studies of volcanoes of differing tectonic types are shown in Table 6. R. B. Symonds and colleagues made several important generalizations based on their summary of recent studies: Subduction volcanoes tend to emit higher H_2O and lower CO_2 and SO_2 relative to rift volcanoes and hot-spot volcanoes (which have high relative CO_2 and SO_2 and low H_2O). Plate tectonics influences these abundances. The subducted slab contributes H_2O and HCl to magmas at convergent margins. Magma sources for rift volcanoes and hot-spot volcanoes are located in the mantle; they contain moderate amounts of CO_2 and SO_2 and are relatively low in H_2O and HCl. Two-stage degassing is probably common at hot-spot and rift volcanoes. For example, Kilauea magmas apparently saturate with CO_2-rich fluid at about 40 kilometers depth, but most of the other gases exsolve at less than 150 meters, making the magmas more H_2O-rich near the surface.

Recent study and analysis of the submarine hot springs that occur along mid-ocean ridges indicate that they too are sources of volcanic gases, especially carbon dioxide. The mid-ocean ridge lavas contain a low

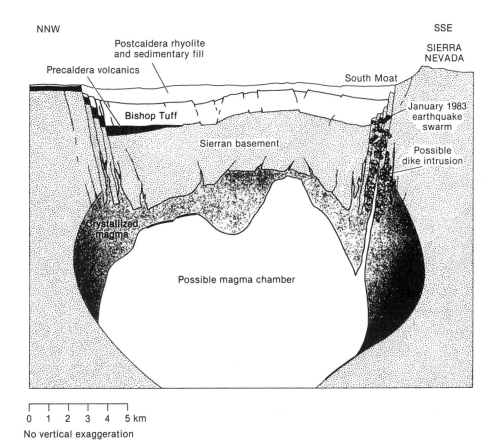

NNW

SSE

SIERRA
NEVADA

Postcaldera rhyolite
and sedimentary fill

Precaldera volcanics

South Moat

Bishop Tuff

January 1983
earthquake
swarm

Sierran basement

Possible
dike intrusion

Crystallized
magma

Possible magma chamber

0   1   2   3   4   5 km

No vertical exaggeration

128 A schematic cross section through Long Valley Caldera near Mammoth Lakes, California. This caldera formed when 600 cubic kilometers of pyroclastic flows called Bishop Tuff were erupted about 700,000 years ago. The most recent eruptions, 500 to 600 years ago, formed a north–south line of rhyolitic domes in and north of the caldera. Earthquake swarms and uplift within the caldera from 1980 to 1996 indicate a probable injection of new magma beneath the area. (Adapted from D. P. Hill, R. A. Bailey, and A. S. Ryall, *Journal of Geophysical Research* 90 [1985]: 11111–11120 and cover.)

percentage of gas, but their abundance compared with the lavas erupted on land makes them a major supplier of gases to the Earth's surface.

The sum of all these studies allows some reasonable estimates of the volume and composition of the volcanic gases currently being added to the Earth's water and air. Based on the numbers of their atoms, hydrogen is the most important constituent of volcanic gases, followed by oxygen, carbon, sulfur, chlorine, and nitrogen. As these elements combine at the surface of the Earth, they become water, carbon dioxide, sulfur dioxide, hydrochloric acid, and nitrogen. The ratios of most of these con-

TABLE 6   Examples of the composition of volcanic gas emissions

Volcano	Kilauea Summit	Erta Ale	Momotombo
Tectonic Style	Hot Spot	Rift	Subduction
Temperature	1170°C	1130°C	820°C
Gas	Percentage of emission by volume		
H_2O	37.1	77.2	97.1
CO_2	48.9	11.3	1.44
SO_2	11.8	8.34	0.50
H_2	0.49	1.39	0.70
CO	1.51	0.44	0.01
H_2S	0.04	0.68	0.23
HCl	0.08	0.42	2.89
HF	—	—	0.26

Source: Symonds et al., 1994.

stituents in volcanic gases are in reasonable accord with the ratios of water, carbon, chlorine, and nitrogen found in the Earth's air, oceans, and surface rocks. However, the ratio of sulfur is not.

Sulfur appears to be a special problem. At its current rate of emission from volcanoes, there should be 50 times more sulfur at the Earth's surface than there is. What has saved us from the sulfuric acid blanket that plagues the planet Venus? The answer is probably iron sulfide—the common mineral pyrite, or "fool's gold"—as well as the plate tectonic process. Instead of being released into the ocean or atmosphere, sulfur reacts with iron and forms insoluble iron sulfide, which remains in the oceanic crust and is returned via subduction to the shallow interior of the Earth.

Estimates of the amounts of water and chlorine that are being subducted are in rough agreement with the release of these volatiles from subduction volcanoes. In addition, the geologic record supports the belief that the volume and salinity of the oceans have not greatly changed over time. These data support the concept that the recycling of some volatiles is in balance. As Paul Wallace and Fred Anderson point out in their 2000 paper, "subduction has not caused the oceans to dry up."

Although the compositional ratios appear generally correct for a partly volcanic origin of the Earth's air and water, what about the total amounts of these gases? Figure 129 shows the volatile elements and compounds in the Earth's oceans, sedimentary rocks, and atmosphere compared with their volcanic production. From an accounting standpoint, there is more

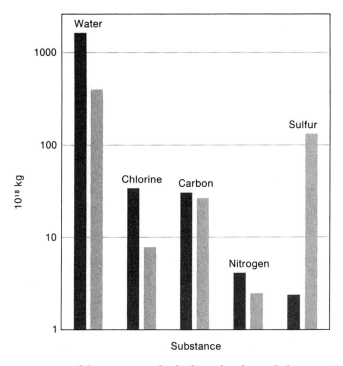

129 A comparison of the amounts of volatile molecules and elements in the Earth's oceans and atmosphere with their volcanic production. (Note the logarithmic scale.) For each substance, the darker bar on the left represents the total weight of that material in the Earth's air, oceans, and sedimentary rocks. The lighter bar on the right represents the weight of that same substance currently being produced from volcanoes multiplied by the Earth's age. It is apparent that some process such as subduction must remove sulfur from the Earth's surface.

water, chlorine, and nitrogen than volcanic exhalations can explain, and there is far less sulfur. Only carbon appears to be in rough balance.

"Appears" is the correct word, for the data in Figure 129 are still being refined; the total abundances of water, chlorine, carbon, and nitrogen are fairly well established, but sulfur in the top layers of the oceanic crust is not included. The volcanic exhalations are only first approximations, and it is not known whether the current volcanic activity is representative of volcanism throughout geologic time. Nevertheless, a comparison based on approximate data can be important for qualitative purposes, even though future data may change its quantitative conclusions. In other words, the current data support the concept of recycling by subduction, even though the total amount of various volatiles that have been recycled is still uncertain.

If the water that still covers most of our planet's surface was not originally part of the Earth, it must have been added early from outside. Comets, made largely of ice, are likely candidates for this extraterrestrial source. Apparently comets still occasionally collide with the Earth, and during the early days of the solar system the number of comets and their collisions is thought to have been much greater.

If comets and not volcanoes provided much of the Earth's early sea and air, volcanologists can no longer claim that these essentials of life were all born of fire. Nevertheless, recycling is an important task, especially if the overall process can clear the air of sulfuric acid. Ongoing volcanism and tectonic processes have helped provide the long-term balance among land, sea, air, and climate that is essential to the evolution of life on Earth.

Links

Degassing Nyos Volcano
http://perso.wanadoo.fr/mhalb/nyos/

Lake Nyos, J. P. Lockwood
http://volcanologist.com/pages/1986.html

Long Valley Caldera, California
http://wrgis.wr.usgs.gov/fact-sheet/fs108-96/

Mammoth CO_2 Tree Kill
http://wrgis.wr.usgs.gov/fact-sheet/fs172-96/

Origin of Earth, Oceans, and Life
http://www.ic.ucsc.edu/~eart1/Notes/Lec1.html

Sulfur in Wikipedia (the free encyclopedia)
http://en.wikipedia.org/wiki/Sulfur

Volcanic Air Pollution
http://wrgis.wr.usgs.gov/fact-sheet/fs169-97/

Volcanic Gases and Their Effects
http://volcanoes.usgs.gov/Hazards/What/VolGas/volgas.html

Volcanic Gases Fact Sheet
http://vulcan.wr.usgs.gov/Projects/Emissions/vgas_fsheet.html

Sources

Arthur, M. A. "Volcanic Contributions to the Carbon and Sulfur Geochemical Cycles and Global Change." In H. Sigurdsson, ed., *Encyclopedia of Volcanoes*. San Diego, CA: Academic Press, 2000.

Butterfield, D. A. "Deep Ocean Hydrothermal Vents." In H. Sigurdsson, ed., *Encyclopedia of Volcanoes*. San Diego, CA: Academic Press, 2000.

Delmelle, P., and J. Stix. "Volcanic Gases." In H. Sigurdsson, ed., *Encyclopedia of Volcanoes*. San Diego, CA: Academic Press, 2000.

Gerlach, T. M. "Present Day CO_2 Gas Emissions from Volcanoes." *Eos* (*Transactions, American Geophysical Union*) 72 (1991): 249, 254–255.

Gerlach, T. M. "Volcanic Gases." *Encyclopedia of Geochemistry*. Encyclopedia of Earth Science Series. London: Chapman & Hall, 1997.

Gerlach, T. M., H. R. Westrich, and R. B. Symonds. "Pre-Eruption Vapor in Magma of the Climactic Pinatubo Eruption." In C. G. Newhall and R. S. Punongbayan, eds., *Fire and Mud: Eruptions and Lahars of Mount Pinatubo, Philippines*. Quezon City: Philippine Institute of Volcanology and Seismology; Seattle: University of Washington Press, 1996.

Graedel, T. E., and P. J. Crutzen. *Atmospheric Change: An Earth System Perspective*. New York: W. H. Freeman, 1995.

Holland, H. D. *The Chemical Evolution of the Atmosphere and Oceans*. Princeton, NJ: Princeton University Press, 1984.

Kasting, J. F. "Earth's Early Atmosphere." *Science* 259 (1993): 920–926.

McGee, K. A., and T. J. Casadevall. "A Compilation of Sulfur Dioxide and Carbon Dioxide Emission-Rate Data from Mount St. Helens during 1980–1988." U.S. Geological Survey Open-File Report 94-212, 1994.

Stoiber, Richard. "Volcanic Gases from Subaerial Volcanoes on Earth." In Thomas Ahrens, ed., *Global Earth Physics*. AGU Reference Shelf, no. 1. Washington, DC: American Geophysical Union, 1995.

Symonds, R. B., W. I. Rose, G. J. S. Bluth, and T. M. Gerlach. "Volcanic Gas Studies." In M. R. Carroll and J. R. Hollaway, eds., *Volatiles in Magma*. Vol. 30 of *Reviews in Mineralogy*. Mineralogical Society of America, 1994.

Wallace, P., and A. T. Anderson. "Volatiles in Magma." In H. Sigurdsson, ed., *Encyclopedia of Volcanoes*. San Diego, CA: Academic Press, 2000.

CHAPTER 15

Volcanic Power

130 The Wairakei geothermal field generates electricity in New Zealand.

Plate 11 The eruption of a geyser in central Iceland. *Geysir*, an Icelandic word meaning "to gush," is the original word for these spectacular spouting hot springs. Other noteworthy examples can be found in Yellowstone National Park and near Rotorua in New Zealand.

Plate 12 Mount St. Helens in 1969, viewed from Spirit Lake. Before the massive eruption of 1980, the summit elevation of this beautiful stratovolcano was 2950 meters.

Plate 13 Mount St. Helens and Spirit Lake in 1982. The rim of the 2-kilometer-wide, horseshoe-shaped crater reaches an elevation of 2549 meters, and the crater bottom is more than 1000 meters below the pre-1980 summit elevation. The new lava dome can be seen inside the crater. (Photograph by Lyn Topinka, U.S. Geological Survey.)

Plate 14 An explosive eruption of Mt. St. Helens on July 22, 1980. Although this explosion generated an ash cloud that reached a height of 14 kilometers, its energy was about a hundred times less than that of the May 18, 1980, eruption. (Photograph by Katia Krafft.)

Plate 15 Renewed eruption of Mount St. Helens on October 1, 2004, after 18 years. A new lava dome pushing up beneath the glacier in the crater caused the small release of steam. The gray area is the old 1980-1986 lava-dome complex. (Photograph by John Pallister, U.S. Geological Survey)

Plate 16 The eruption cloud from Mount Pinatubo in the Philippines on June 12, 1991. The climactic eruption occurred on June 15, 1991. (Photograph by David Harlow, U.S. Geological Survey.)

Plate 17 Mount Rainier, a 4393-meter-high, dormant stratovolcano in Washington State. The 25 glaciers on Rainier contain 4 cubic kilometers of ice that could cause a debris avalanche and mudflow if an eruption or large earthquake should take place.

Plate 18   A fumarole (volcanic gas vent) in Kilauea Caldera deposits sulfur as the gases cool and mix with air. (Photograph by R. L. Christiansen, U.S. Geological Survey.)

Plate 19   Submarine hot springs have been discovered at many locations along the volcanic mid-ocean ridges. These "black smokers" at 13° north latitude on the East Pacific Rise are pouring 350°C mineral-laden water into the cold, deep bottom water 2500 meters below sea level. Iron sulfide and other metallic sulfide particles precipitating from the suddenly chilled hot-spring water form the black "smoke." (Photograph by J. L. Cheminée, Observatoires Volcanologiques, Institut de Physique du Globe de Paris, doc. Ifremer; taken from the French submersible Cyana in 1982.)

Plate 20 Cinder and lava eruption of Mount Etna in Sicily in October 2002. (Copyright photo by Tom Pfeiffer.)

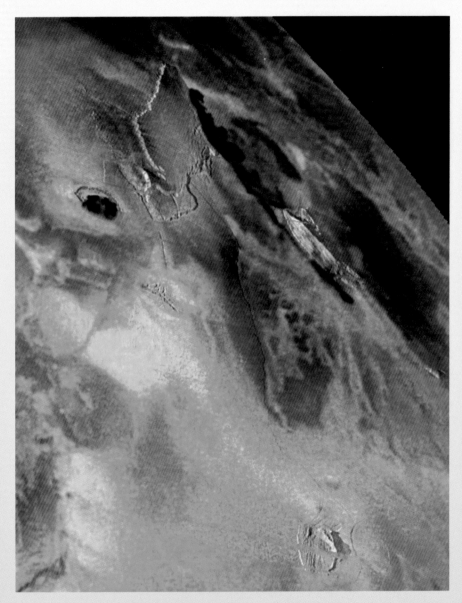

Plate 21 Volcanic deposits from the vigorously erupting volcanoes on Io, a moon of Jupiter, rapidly change its colorful face as seen on images from passing spacecraft. The region of young, dark lava flows visible at the upper right is called Zal Patera. (Image by NASA/JPL.)

> *Fires that shook me once, but now to silent ashes fall'n away*
> *Cold upon the dead volcano sleeps the gleam of dying day.*
>
> —Tennyson (1809–1892)

G*eothermal energy* was a term known to only a few specialists before the 1974 energy crisis, a crisis that has now been almost forgotten. Although not exactly a household word, it has become a more familiar term as discussions of alternative sources of energy have increased.

In discussing geothermal energy and power, it is important to understand the subtle difference between these terms. Heat is a form of *energy*, and there is an immense amount of it inside the Earth. For each kilometer of depth, the temperature increases about 20°C to 60°C, depending on the region (Figure 131). The heat energy contained in just the upper 10 kilometers of the United States is estimated to be 3.3×10^{25} joules, equal to the heat released by burning a thousand trillion tons of coal. If it were accessible, it would supply our energy needs for the next 100,000 years.

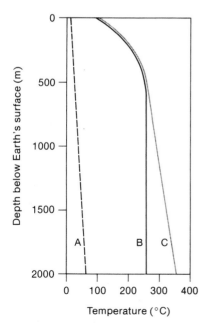

131 Temperatures generally increase with depth beneath the Earth's surface. Temperatures in regions of hot springs and geysers (curve B) are distinctly different from those in nonvolcanic regions (curve A). Curve C is the temperature of boiling water under theoretical hydrostatic pressure found at various depths. (Adapted from D. E. White, U.S. Geological Survey.)

But potential energy becomes practical only when it can be consumed at some useful rate. *Power* is a measure of the rate at which energy is made available or is consumed. Energy is measured in joules, and power is measured in joules per second, or watts. There is plenty of diffuse energy in the Earth; turning it into useful power is the critical problem.

The difference between energy and power becomes clear in the following examples: The amount of solar energy is enormous, but the solar power reaching the Earth is modest—averaging about 40 watts for each square meter of surface. A lightning bolt contains only a modest amount of energy, equal to that of about two barrels of crude oil, but it expends this energy in a fraction of a second, unleashing enormous power.

Although volcanic explosions are often compared with nuclear bomb blasts, the analogy is misleading. The heat and mechanical energy in the 1991 eruption of Mount Pinatubo was on the order of 1500 megatons of TNT, but this energy was released in a series of explosions lasting almost a day. A nuclear bomb, however, goes off in a single, nearly instantaneous flash, unleashing power in amounts unmatched by geologic phenomena.

Geothermal energy—like solar energy—exists in enormous amounts, but its natural rate of release is small, averaging about one-sixteenth of a watt from beneath each square meter of the Earth's surface. Even if geothermal energy could be converted to electricity with an efficiency of 20 percent, it would require all the heat flow from an area as large as a football field to power one 60-watt lightbulb. Only in a few volcanic regions such as Yellowstone, New Zealand, and Iceland, where geysers and hot springs are abundant, does nature sustain a significant output of geothermal power (Figure 130).

For geothermal power to be practical, some special situation must exist that can concentrate the Earth's heat energy into a small area. Natural or artificial underground reservoirs of steam or hot water that can be funneled into a drill hole provide this special situation (Figure 132). In nonvolcanic areas, holes drilled to depths of 4 or 5 kilometers may reach temperatures near 100°C. However, in young volcanic regions, where molten rock has brought heat up from deeper levels, it is possible to drill into rocks and hot water or steam reservoirs at depths of 1 to 3 kilometers that are heated to 100°C–350°C or hotter. To develop the true potential of geothermal power, it is necessary to locate these shallow bodies of hot rocks and fluids and to tap their pent-up energy through a system of drill holes. Exploring for geothermal power is similar to exploring for oil, except that the geologic setting is volcanic rather than

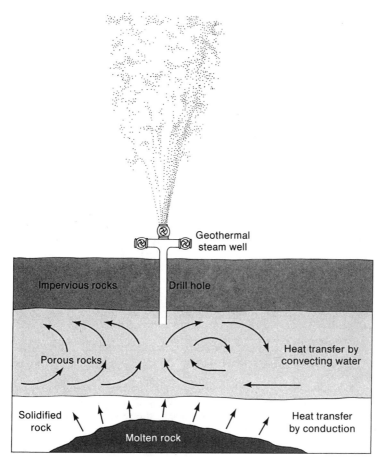

132 A schematic diagram of a geothermal reservoir.

sedimentary. Most of the existing high-temperature geothermal power fields have been found in active or young volcanic regions.

The form of heat transfer in a given region is of key importance. Heat is transferred by radiation, conduction, and convection. *Radiation* transfers a significant amount of heat through transparent media such as air and space, but it is much less effective in solids. *Conduction* is such a slow process in good thermal insulators like rocks that it transfers only small amounts of heat from inside the Earth. The very slow release of the Earth's interior heat through its surface (averaging 0.06 watt per square meter) is controlled by this slow conduction process. *Convection* is the only process fast enough to transfer the Earth's heat at rates sufficient to produce significant power. Convection involves the actual physical movement of molten rock or hot fluids, generally upward, because

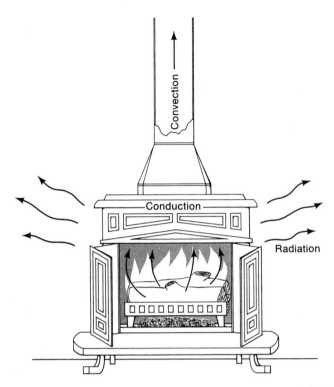

133 A sketch of a woodstove shows heat transfer by convection of the heated air and wood smoke in the stove; by conduction through the firebrick and iron walls of the stove; and by radiation from the surface of the stove through the open space of the room.

these hot rocks and fluids are less dense than the cooler rocks and fluids surrounding them (Figure 133).

Geothermal power fields are found where magma has moved upward from depths of 50–100 kilometers and brought the high temperatures (900°C–1200°C) of these depths to near the surface (Figure 134). Groundwater heated by these volcanic intrusions can form another convection circuit, bringing hot springs and geysers to the surface or remaining sealed below the surface awaiting the wildcatter's drill.

Geothermal prospects related to volcanic activity are of three types: hydrothermal reservoirs, hot dry rocks, and magma reservoirs. Of the three, hydrothermal reservoirs are the easiest to develop because their hot water and steam are waiting to be tapped.

Larderello, Italy, is an area of natural hot springs. The boric acid concentrated in these thermal waters has been extracted for more than 200 years. Wells drilled to increase chemical production encountered steam and were first used to produce electricity in 1904. Some 300 wells have

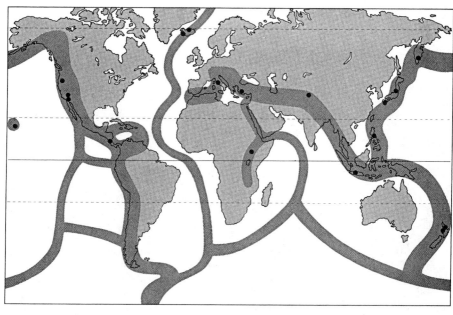

Key

▬ Regions favorable for geothermal prospecting

● Location of geothermal power plants

134 A map of the world's geothermal regions. Notice the close relationship to zones of active volcanoes.

been drilled to depths of 300 to 700 meters over the years and have outlined a reservoir producing steam at 235°C and 30 bars of pressure. The reservoir is in porous limestone, but intrusions of young volcanic rocks in the area are considered to be the heat source. The field has produced some 400 megawatts of power for the last 50 years and is still going strong (Figure 135).

In 1925, Icelanders began to drill for hot water and steam. By now they have developed more than 250 fields whose power is used mainly for space heating. Almost all of Reykjavík, the capital city of nearly 150,000 people, gets its heat and hot water from more than 100 wells, 300–2200 meters deep, that tap water at about 100°C in porous basalts. Although these once-artesian wells now need to be pumped, the temperatures have shown little decrease. For the people in oil-barren, arctic Iceland, geothermal power is the very lifeblood of their warmth and comfort (Figure 136).

In 1921, The Geysers in California—a region of young volcanic rocks in the Coast Range north of San Francisco—was drilled for steam. Electricity was first generated in 1959, and the field has been greatly expanded

135 Larderello, Italy, is the site of the world's first geothermal steam field. First developed in 1904, it now produces about 300 to 400 megawatts of electricity. The huge chimneys are cooling towers similar to those used at nuclear power plants for recycling the cooling water. (Photograph by Patrick Muffler, U.S. Geological Survey.)

since that time. More than 600 wells, 500–3200 meters deep, once produced steam at about 240°C and 30 bars of pressure from a reservoir of fractured shaley sandstone, volcanics, and intrusive rocks. Production was expanded during the 1980s to nearly 2000 megawatts, enough electricity for a city of 2 million people (Figure 137). In recent years, however, production has declined, down to 1200 megawatts by 1993. Once considered a nearly inexhaustible source of power, the field has lost pressure as a result of the rapid extraction of huge amounts of steam. Even so, the 1000 megawatts of electricity currently generated at The Geysers makes it the world's largest geothermal power field, and new ideas for water injection and management of steam production rates will probably extend its useful life for several decades.

On the island of Hawaii, a geothermal exploration well encountered some of the hottest underground fluids found so far. Drilled into lava flows on the east rift of Kilauea Volcano to a depth of 1969 meters, the well had a bottom temperature of 350°C. This discovery well generated

136 *Geysir* is an Icelandic word meaning "to gush," but the Great Geysir of Haukadalur, the world's original geyser, seldom performs. Its small neighbor, Strokkur Geysir, seen in this photograph, erupts every few minutes.

about 2 to 3 megawatts of electricity during the 1980s (Figure 138). Additional wells have been drilled nearby, and a 25-megawatt generating plant is currently producing a significant part of the electricity consumed on the Big Island.

The prospects for generating geothermal power from hot dry rocks are more speculative. These potential fields are areas of higher-than-normal thermal gradients on the order of 60°C per kilometer of depth in areas of impermeable rocks. Holes drilled to depths of 3 or 4 kilometers can reach down into these hot dry rocks. Artificial fractures are established between two closely spaced holes, and cold surface water is

137 The Geysers, California, once a hot-spring health spa, is now the world's largest geothermal steam field. Producing nearly 1000 megawatts, the field generates enough electricity to supply a city of 1 million people. (Photograph by Pacific Gas and Electric Company.)

pumped down an injection well. Heated by the hot rock along the fractures, this water returns to the surface as hot water or steam in a production well (Figure 139).

After the energy crisis of the mid-1970s, the Los Alamos National Laboratory in New Mexico performed a hot-dry-rock experiment on the flank of the Valles Caldera during the 1980s. Cold water pumped down a 3-kilometer-deep well returned at 135°C up another well close by. However, the cost of the experiment far exceeded the value of the power produced. Its proponents declared the experiment a scientific success, and its opponents declared it a commercial failure. Further experiments have been conducted or are under way in Europe, Japan, Australia, and the United States.

As long as relatively inexpensive oil and natural gas are available, the concept of power from hot dry rock (often referred to as "Engineered Geothermal Systems") will remain an idea whose time has not yet come. Nevertheless, its potential is enormous. Much of the young mountain terrain in the western United States, as well as that in Hawaii and Alaska,

138 Hawaii's first geothermal steam well is located on the east rift of Kilauea Volcano. A 3-megawatt electrical generator was installed on this well. (Photograph by Larry Kadooka, *Hawaii Tribune Herald.*)

is of volcanic origin and constitutes a tempting but well-buried treasure of geothermal energy.

Even molten rock itself is a potential source of power. Magma contains about 1000 joules of heat energy per gram. The energy in 1 cubic kilometer of magma is enough to light San Francisco for 200 years. But tapping it requires truly advanced technology: although scientifically it is feasible to extract power directly from shallow bodies of magma beneath volcanic centers, the engineering technology is still years away.

Initial experiments that involved drilling into buried masses of molten rock were conducted on a lava lake in Kilauea Iki Crater in Hawaii. This great lake of lava formed during an eruption of Kilauea Volcano in 1959, when lava flows ponded in an ancient crater to a depth of 100 meters. A solid crust soon formed on the lake, but lava is such a good insulator that in 1979, holes drilled deeper than 50 meters were still reaching partially molten rock. Conventional rock-drilling techniques using abundant water to keep the drill bit cool were successful in reaching the molten rock, although efforts to place devices down the holes into the molten rock to determine the feasibility of extracting energy were not successful.

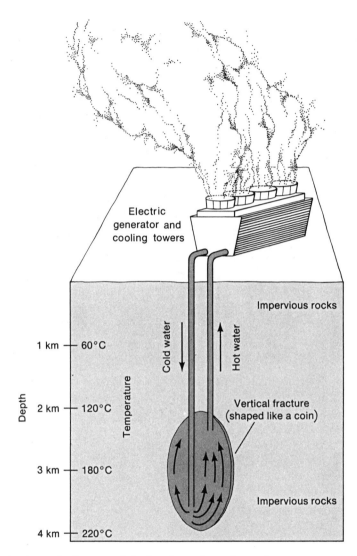

139   A schematic diagram of the hot-dry-rock geothermal power concept. Two holes are drilled into impervious rocks in a region where the temperature increases rapidly with depth. The holes are connected by hydraulic fracturing, a process of fracturing rock by pumping fluid at high pressure down a well. Surface water is injected into one hole; it is heated by contact with the hot rocks at depth and rises in the second hole as hot water or steam. Cooling of the rocks at depth causes further fracturing, allowing the circulating water to reach additional hot surfaces. Preliminary tests near Los Alamos, New Mexico, with wells drilled to a depth of 3 kilometers in granite at 200°C, have been encouraging.

Even though geothermal power is still a young industry, its potential makes it worth serious effort and investment. In an assessment of geothermal power in the United States, the U.S. Geological Survey estimated that identified resources of geothermal reservoirs (excluding national parks) exceeding 150°C are sufficient to generate 23,000 megawatts of electricity for 30 years, and that identified geothermal reservoirs of less than 150°C are sufficient to generate 280,000 megawatts of thermal power for hot water and heating for 30 years. Most geothermal fields have been discovered by drilling where there are surface hydrothermal features such as hot springs, but the number of hidden fields is difficult to estimate. Table 7 lists the geothermal power production of nations worldwide as of 2004.

TABLE 7   **Geothermal power production 2001–2005**

Country	Megawatts of Electricity
Australia	0.1
China	19
Costa Rica	163
El Salvador	119
Ethiopia	7
France	15
Germany	2
Guatemala	29
Iceland	202
Indonesia	838
Italy	699
Japan	530
Kenya	127
Mexico	953
New Zealand	403
Nicaragua	38
Papua New Guinea	6
Philippines	1838
Portugal (Azores)	13
Russia	79
Thailand	3
Turkey	18
United States	1914
Total	8010

Source: Ruggero Bertani, "World Geothermal Generation 2001–2005," Proceedings of World Geothermal Congress, Antalya, Turkey, 2005.

140 Old Faithful geyser in Yellowstone National Park has erupted every 40 to 80 minutes for the past 100 years. Spraying 50,000 kilograms of boiling water 35 to 50 meters high, each eruption releases about 1.9×108 joules of energy during its few minutes duration (about 1 megawatt of power).

Another asset of geothermal power is that it is relatively free of pollution compared with the burning of fossil fuels. Geothermal steam and hot water do not produce soot and usually contain only small amounts of carbon dioxide, one of the "greenhouse" gases that may cause global warming. As world population grows and world energy resources decline, the geothermal power industry will certainly become an important component in sustaining civilization as we know it.

Volcanoes, directly and indirectly, are powerhouses of enormous potential. The most scenic, such as those in Yellowstone, Hawaii Volcanoes, and other volcanic national parks, should never be developed (Figure 140). It is possible that other volcanoes will help provide us with light and warmth in centuries to come, but we will have to undertake a major research and engineering effort, involving much time, labor, and money, to find out. As with all human adventures, high payoff involves high risk.

Links

California Geothermal
http://www.geothermal.org/articles/California.pdf

Geothermal Education Office
http://geothermal.marin.org/

Geothermal Energy Brochure
http://egi-geothermal.org/GeothermalBrochure.pdf

Geothermal Networks
http://www.geothermie.de/index.html

Geothermal Resources Council
http://www.geothermal.org/

International Geothermal Association
http://iga.igg.cnr.it/geoworld/geoworld.php

Lassen Hydrothermal Features
http://geopubs.wr.usgs.gov/fact-sheet/fs101-02/

Long Valley Drilling
http://www.sandia.gov/media/mammouth.htm

Sources

Arnorsson, Stefan. "Exploitation of Geothermal Resources." In H. Sigurdsson, ed., *Encyclopedia of Volcanoes*. San Diego, CA: Academic Press, 2000.

Bertani, Ruggero. "World Geothermal Generation 2001–2005." Proceedings of World Geothermal Congress, Antalya, Turkey, 2005.

Cataldi, R., S. F. Hodgson, and J. W. Lund, eds. *Stories from a Heated Earth: Our Geothermal Heritage*. Davis, CA: Geothermal Resources Council, 1999.

Duffield, Wendell, J. H. Sass, and M. L. Sorey. *Tapping the Earth's Natural Heat*. U.S. Geological Survey Circular 1125, 1994.

Goeff, F., and C. J. Janik. "Geothermal Systems." In H. Sigurdsson, ed., *Encyclopedia of Volcanoes*. San Diego, CA: Academic Press, 2000.

Hochstein, M. P., and P. R. L. Browne. "Surface Manifestations of Geothermal Systems with Volcanic Heat Sources." In H. Sigurdsson, ed., *Encyclopedia of Volcanoes*. San Diego, CA: Academic Press, 2000.

Lovekin, J. W. "Sustainable Geothermal Power: The Life Cycle of a Geothermal Field." *Geothermal Resources Council Transactions* 22 (1998): 515–519.

Muffler, L. J. P., ed. *Assessment of Geothermal Resources of the United States—1978*. U.S. Geological Survey Circular 790, 1979.

Rybach, L., and L. J. P. Muffler. *Geothermal Systems: Principles and Case Histories*. New York: Wiley, 1981.

Wohletz, K., and G. Heiken. *Volcanology and Geothermal Energy*. Berkeley and Los Angeles: University of California Press, 1992.

CHAPTER 16

Volcanic Treasures

141   Bingham Canyon, Utah, open-pit porphyry copper mine. (Courtesy of Rio Tinto Limited.)

Volcanoes are nature's forges and stills where the elements of the Earth, both rare and common, are moved and sorted. Some elements are diluted and some pass through unchanged, but many are transported and concentrated into those precious lodes that people seek for fortune or industry.

Economically important elements concentrated directly by volcanic action, especially by the intrusion of magma bodies into the Earth's crust, include fluorine, sulfur, zinc, copper, lead, arsenic, tin, molybdenum, uranium, tungsten, silver, mercury, and gold. These concentrations are mainly formed as hydrothermal vein deposits, which are precipitated from hot waters percolating along underground fractures. Such veins generally consist of one or more common minerals, like quartz or calcite (calcium carbonate), within which the more precious minerals, such as gold or galena (lead sulfide), are scattered as small specks or crystals.

Theoretically the concentration process is simple, but in operation it can be enormously complex. Basically, the magmatic roots of volcanoes supply the heat source, and perhaps some of the ingredients, for a giant still. As the magma cools and the common silicate minerals crystallize to form gabbro or granite, the water and other gases—as well as the rarer elements that do not fit into the rock-forming silicate minerals—become concentrated in the residual silicate melt. As cooling reduces the volume of the rocks, they crack, allowing the hot residual magmatic fluids, rich in water and precious elements, to escape from their underground forge and ascend toward the surface (Figure 142). During their ascent, cooling and decreasing pressure cause various minerals, both common and rare, to precipitate from these hydrothermal solutions and form veins or disseminated deposits. Certain minerals precipitate over a large range of pressures and temperatures and are common throughout a deposit; others, such as gold and silver, may precipitate over a narrow range of pressures and temperatures to form localized bonanzas.

Veins are aggregates of minerals found in rock fractures, often occurring as steeply inclined planes dipping into the Earth, a few centimeters to many meters thick and often many hundreds of meters or several kilometers in length. The silver-bearing vein of the Comstock Lode near

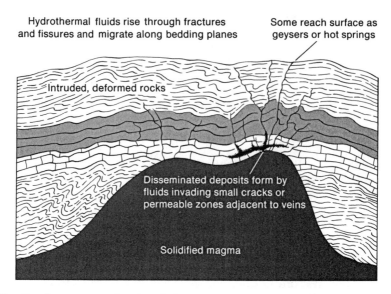

Hydrothermal fluids rise through fractures and fissures and migrate along bedding planes

Some reach surface as geysers or hot springs

Intruded, deformed rocks

Disseminated deposits form by fluids invading small cracks or permeable zones adjacent to veins

Solidified magma

142 Ore deposits are often found close to an intrusion of solidified magma. Many such deposits probably precipitate out of the hot waters associated with the cooling magma. (From F. Press and R. Siever, eds., *Earth*, 2nd ed., p. 577. Copyright © 1978, W. H. Freeman and Company.)

Reno, Nevada, trends north–south and dips steeply down toward the volcanic roots of the Virginia Range. The hydrothermal origin of the Comstock Lode minerals was still evident in the high temperatures encountered by the miners who rushed to Nevada in the 1860s to seek their fortunes in silver.

Disseminated deposits of copper minerals, sometimes with significant gold and molybdenum content, often occur within the shallow magmatic roots of volcanoes. These deposits of scattered ore minerals are quite low grade—roughly only 1 percent copper—but they occur in such large volumes that they can be mined in vast open-pit operations. Bingham Canyon in Utah and Chuquicamata in Chile are classic examples of this variety of ore deposit associated with volcanism.

The type of rock surrounding the cooling magma reservoir is important to the formation of ore deposits. In Hawaii (and, in fact, on most basaltic oceanic volcanic islands), ore deposits are nearly nonexistent. Part of the reason for this may be that most oceanic islands are slowly sinking under their own enormous load, thus hiding any ore deposits that have formed. But even in the unusual cases in which uplift and erosion have exposed the roots of these volcanic islands, hydrothermal ore deposits are rare or absent. Volcanic rocks richer in silica than basalt—for example, andesite, dacite, and rhyolite—are more often associated with hydrothermal ore deposits. Many geologists believe that the rocks surrounding the magmatic roots of volcanoes are the true source of the valuable elements found in hydrothermal veins. According to this view, the volcanic rocks act as a heat source that pumps existing groundwater into a giant circulating system. Cold waters, being heavier, move down and into the cooling volcanic rocks, carrying trace quantities of valuable elements leached from the surrounding rocks. Heated by the cooling magma, they become less dense and rise into the fractured rocks above. Cooling and losing pressure again, the waters precipitate their quartz and precious ores into the veins forming above the volcanic hearth (Figure 143).

Because mercury, copper, sulfur, and fluorine have been measured in volcanic gases in Hawaii, there is no question that at least some of these elements originate directly in volcanoes. Gold, silver, and other valuable elements found in hydrothermal ore deposits may have to have been concentrated beforehand in some earlier generation of continental crustal rocks before they can be reconcentrated by volcanic distillation.

Two interesting discoveries about the relation of gold to subduction volcanoes have been reported. Geologists studying the fumarole and hot-

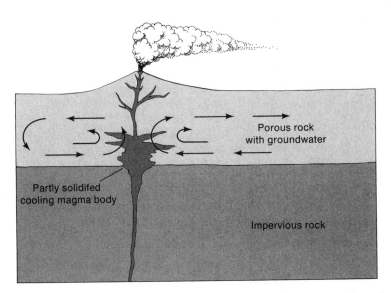

143 A schematic diagram of groundwater being circulated by the cooling of a magma body. The water seeps into the cooling magma body, is heated, and convects upward. Reaching cooler rocks above the magma body, the circulating groundwater also cools and precipitates vein minerals. The source of the minerals may be either the magma or the porous rock through which the groundwater circulates. In this model, the cooling magma body acts as a circulating pump and concentrator rather than as a direct source of ore deposits.

spring chemistry of Galeras Volcano in Colombia have found that 0.5 kilogram of gold is escaping to the atmosphere every day from the gas vents, and that more than 20 kilograms of gold per year are being deposited inside the volcano. Of course these emanations and deposits are very low grade, but if all the gold could somehow be collected, it would amount to more than $2 million worth per year.

The other startling find is a major gold discovery that apparently is still being deposited by a shallow, active geothermal system on Lihir, a volcanic island in Papua New Guinea. The deposit lies in a caldera on this young volcano. Speaking in geologic time, of course, it has been said that today's ore deposits were often yesterday's geothermal systems, but here is a place where both are present today. It is ironic that in order to mine the gold at Lihir, the geothermal system will somehow have to be cooled first.

Each mineral has its characteristic place in the hydrothermal system. Tungsten minerals precipitate at very high temperatures and often are found at the very point of contact between a chilled magma body and the rocks it has invaded, especially limestone. Scheelite, a calcium tungstate mineral, is brightly fluorescent under ultraviolet light. Tungsten prospectors seek out contacts between granite and limestone during the day and then prospect these outcrops at night with "black lights" (ultraviolet lamps) to reveal the areas of high tungsten concentration.

Mercury is at the other extreme. Both as an element and in compounds, it is volatile. For this reason, much of the Earth's mercury is lost to the surface from volcanic gas vents and hot springs. Near many active volcanoes the concentration of mercury in the air exceeds the health standards established by environmental protection agencies, but it is difficult to get the volcanoes to stop smoking.

Hydrothermal ore deposits in continental locations have been studied by geologists for more than a hundred years, and were mined for thousands of years before that, but only recently has the importance of submarine ore bodies deposited at mid-ocean rifts been recognized. In the 1960s scientists probing the floor of the Red Sea discovered some areas where hot but dense, highly saline bottom water overlies muds that are rich in iron, zinc, and copper minerals. Since the Red Sea is a volcanic rift zone, oceanographers suspected that the metal-rich sediments might have been precipitated from submarine hot springs.

Proof of this idea came in the late 1970s, when scientists in submersibles peered out on the strange landscapes of volcanic rift zones in the eastern Pacific Ocean. At 2500 meters below the surface, they were startled to see the "black smokers" that we described in Chapter 3 — chimneys of iron, zinc, and copper sulfides a few meters high, pouring out dark particles of iron sulfides. The water spouting from these chimneys is hot—350°C—and as it is quickly chilled by the surrounding seawater, the dissolved metal sulfides precipitate into fine particles (the black smoke) or into the chimneys. Some of these metal sulfide deposits contain small but significant amounts of silver and gold. In addition to building the chimneys, the ore deposits form mats and crusts as much as a few meters thick that are spread over several hundred square meters around the vents.

The general model for the operation of these submarine hot springs and their attendant ore deposits is shown in Figure 144. Heat from the magma chamber beneath the rift zone drives a convective circulation of seawater within the porous basalts of the seafloor. Water sucked in on the rift flanks is heated and dissolves calcium, silica, and metallic sulfides while depositing magnesium and sulfate. The heated discharge waters—containing volcanic fluids rich in carbon dioxide and hydrogen sulfide—are cooled as they emerge on the seafloor and deposit their metal sulfides.

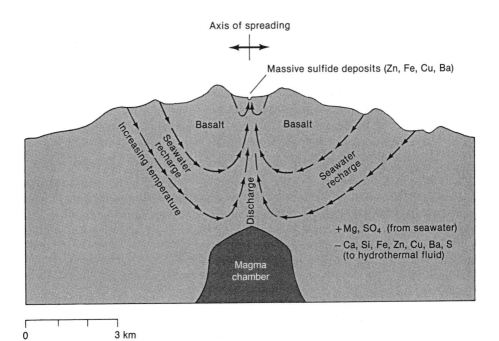

144 A schematic cross section of fluid circulation and hydrothermal deposits at a submarine rift zone. The seafloor topography is exaggerated, and the depth to the magma chamber is uncertain. (R. A. Koski, U.S. Geological Survey.)

The large extent of submarine ore deposits is just now being recognized. They apparently occur in many places along mid-ocean ridges, and as the seafloor spreads away from the ridge, the areas of ore bodies increase. Although most of them are currently so deep that they are out of reach of commercial mining, a few ancient deposits have been shoved up on land where oceanic plates have converged on continents. The metal sulfide mines in Cyprus and Japan, long puzzling to geologists, appear to have originated at submarine hot springs. Most ancient submarine ore deposits, however, probably have been consumed in subduction zones as the ocean plates bearing them pushed beneath a continent. Perhaps they are still being recycled by subduction volcanoes into the more conventional hydrothermal ore deposits found in continental rocks.

Diamonds, perhaps the greatest mineral treasure of all, are also closely related to volcanic processes. Diamonds and the graphite cores of common pencils are the same element—carbon—but they are different minerals. In graphite, the carbon atoms are arranged in layers like the silica atoms in sheets of mica, and the loose bonding between layers permits the sheets to break and slide past one another. Graphite is therefore soft and greasy feeling. In diamonds, the carbon atoms are compressed into a tight network that interlocks in all directions, forming the hardest substance known on Earth (Figure 145). To achieve this close packing of carbon atoms, extremely high pressures are needed—pressures that occur naturally only at depths of nearly 200 kilometers inside the Earth. Once formed, diamonds are stable at low pressures and temperatures, but at the Earth's surface they can burn at high temperatures.

The right conditions to form diamonds apparently exist beneath the continents at a depth of nearly 200 kilometers and at a temperature near that of molten rock (see Figure 120). In fact, diamonds might not be all that rare if we could mine them at their deep source. Their occurrence at the Earth's surface results from a rare type of volcanic eruption that transports them rapidly from great depths into shallow vents called *Kimberlite pipes* (Figure 146). The pressure and temperature are reduced so quickly that the diamonds do not revert to a more normal surface form of carbon, such as graphite.

No one knows whether the type of volcano associated with diamond pipes exists today. The peculiar Kimberlite volcanic rock and the presence of other minerals indicating that the volcanic fluids were extremely high in carbon dioxide suggest an unusual kind of explosive volcanism that may occur only during certain episodes of geologic history. A few active volcanoes in East Africa, especially those whose rocks are unusually high in sodium carbonate (like Ol Doinyo Lengai), may be the closest active relatives of a diamond-pipe volcano (Figure 147).

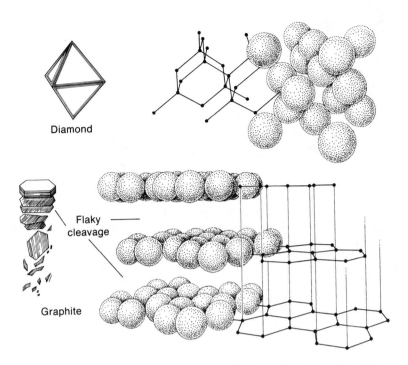

Diamond

Flaky cleavage

Graphite

145 The crystal form and internal structure of diamonds and graphite. The three-dimensional network of atomic bonds in a diamond makes it extremely hard and durable. In contrast, the two-dimensional network of major bonds in graphite makes it flaky and soft. (From J. Gilluly, A. C. Waters, and A. O. Woodford, *Principles of Geology*, 4th ed., p. 30. Copyright © 1975, W. H. Freeman and Company.)

On an expedition to the summit of Ol Doinyo Lengai, scientists found small, active lava lakes in the crater. The temperatures of these molten lavas ranged from 500°C to 550°C, and they appeared to have very low viscosity. The exceptionally low temperatures—about half that of basaltic lavas—apparently result from the high sodium carbonate content of the Ol Doinyo Lengai lavas. The soda acts as a flux, reducing the melting temperature in the same way it does in the manufacture of commercial bottle glass.

Even if they are not transporting diamonds, most of the world's active volcanoes are probably forming some type of hydrothermal deposits beneath their surface today. This is also true of the geothermal reservoirs in which the hot water circulating underground is selectively dissolving, transporting, and depositing the more soluble minerals within the porous rocks. On a geologic time scale, today's volcanoes and geothermal systems are tomorrow's ore deposits.

A

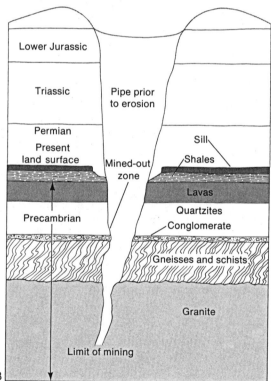

Lower Jurassic

Triassic

Pipe prior
to erosion

Permian

Present
land surface

Mined-out
zone

Sill

Shales

Lavas

Quartzites

Conglomerate

Precambrian

Gneisses and schists

Granite

Limit of mining

B

146 A. The famous Kimberley Diamond Pipe in South Africa. This volcanic vent was mined more than 1000 meters deep before 1908. (Photograph by DeBeers Consolidated Mines Limited.) B. The cross section shows a reconstruction of the pipe before it eroded, the present ground surface, and the mined-out zone. (Diagram after Arthur Holmes.)

147 Ol Doinyo Lengai, "Mountain of God," in Tanzania, East Africa. The high content of sodium carbonate in lavas from this volcano is unique, suggesting that it may be related to the type of volcanism that forms diamond pipes. (Photograph by Richard Stoiber.)

The fact that a natural geothermal area like Yellowstone has a life span of tens of thousands to hundreds of thousands of years indicates that volcanic ore deposits form quite slowly—drop by drop, atom by atom. People and their machines have voracious appetites for minerals, consuming them at rates far in excess of nature's patient creation. The message here is clear: we must recycle whatever mineral wealth we can or wait eons for new supplies.

Links

Argyle Diamond Pipe, Australia
http://volcano.und.nodak.edu/vwdocs/volc_images/australia/argyle/argyle.html

Chuquicamata Minerals
http://www.mindat.org/gallery.php?loc=641

Comstock Lode
http://www.calliope.org/gold/gold3.html

Gold Institute
http://www.goldinstitute.org/

International Association on the Genesis of Ore Deposits
http://nts2.cgu.cz/servlet/page?_pageid=540,542,544&_dad=portal30&_schema=PORTAL30

Nature of Diamonds (American Museum Natural History)
http://www.amnh.org/exhibitions/diamonds/

Ol Doinyo Lengai
  http://magma.nationalgeographic.com/ngm/0301/feature2/index.html
Silver Institute
  http://www.silverinstitute.org/

Sources

Boyd, F. R., and Henry O. A. Meyer, eds. *Kimberlites, Diatremes, and Diamonds: Their Geology, Petrology, and Geochemistry.* Washington, DC: American Geophysical Union, 1979.

Butterfield, D. A. "Deep Ocean Hydrothermal Vents." In H. Sigurdsson, ed., *Encyclopedia of Volcanoes.* San Diego, CA: Academic Press, 2000.

Dilles, J. H., and M. T. Einaudi. Wall-rock alteration and hydrothermal flow paths about the Ann-Mason porphyry copper deposit, Nevada: A 6-km vertical reconstruction. *Economic Geology* 87 (1992): 1963–2001.

Goff, Fraser, J. A. Stimac, A. C. L. Larocque, J. B. Hulen, G. M. McMurty, A. L. Adams, A. Roldan-M., P. E. Trujillo, Jr., D. Counce, S. J. Chipera, D. Mann, and M. Heizer. "Gold Degassing and Deposition at Galeras Volcano, Colombia." *GSA Today* 4, no. 10 (1994): 241–247.

Herzig, Peter, M. Hannington, B. McInnes, P. Stoffers, H. Villinger, R. Seifert, R. Binns, and T. Liebe. "Submarine Volcanism and Hydrothermal Venting Studied in Papua New Guinea." *Eos* (*Transactions, American Geophysical Union*) 75 (1994): 513–516.

Humphris, Susan, R. A. Zierenberg, L. S. Mullineaux, and R. E. Thomson, eds. *Seafloor Hydrothermal Systems.* Geophysical Monograph no. 91. Washington, DC: American Geophysical Union, 1995.

Kesler, S. E., *Mineral Resources, Economics and the Environment.* New York: Macmillan, 1994.

Lipman, P. W. "Ash-flow calderas as structural controls of ore deposits: Recent work and future problems." U.S. Geological Survey Bulletin, 2012-L, L1–L12, 1992.

White, N. C., and R. J. Herrington. "Mineral Deposits Associated with Volcanism." In H. Sigurdsson, ed., *Encyclopedia of Volcanoes.* San Diego, CA: Academic Press, 2000.

CHAPTER 17

Volcanoes and Climate

148 Eruption of Vesuvius, Italy, 1944. (Photograph by the U.S. Navy, courtesy of the National Archives.)

*In one period we believe ourselves governed by
immutable laws; in the next by chance.*
—Loren Eiseley (1907–1977)

Bad weather has been blamed on almost everything—from atomic bombs, sunspots, and the Industrial Revolution to volcanoes, black magic, the Republicans, and the Democrats. And even though most bad weather probably results from the natural variability of atmospheric processes, at least two of these culprits—the Industrial Revolution and volcanoes—have put large amounts of debris and gas into the atmosphere and thus appear more suspect than some of the others (Figure 148).

Benjamin Franklin was the first to suggest that volcanoes modify climate and weather. When the Laki fissure in Iceland (Figure 149) erupted in 1783 with the largest effusion of lava in recorded history, an enormous amount of gas was released, enveloping Iceland and much of northern Europe for months in a blue haze or "dry fog." The gas must have contained a significant amount of fluorine because livestock grazing on contaminated grass in Iceland died of fluoridosis. The widespread death of livestock—11,000 cattle, 28,000 horses, and 190,000 sheep—resulted in a severe famine in which 10,000 Icelanders, one-fifth of the population, perished. By the time the "dry fog" reached Europe it was more annoying than poisonous, but it was visible on many days during the summer and fall of 1783 and was apparently observed by Franklin during his stay in France. Since the winter of 1783–1784 was abnormally severe, especially in Europe, Franklin suggested that the fine ash and gases from the Laki eruption may have filtered out enough of the sun's rays to cause the cold weather.

In 1815 Tambora Volcano shook the island of Sumbawa in Indonesia with a gigantic explosive eruption. Recent expeditions to Tambora estimated that the amount of magma expelled in high ash clouds and pyroclastic flows during that great eruption amounted to about 40 cubic kilometers. Although there are few data on the actual atmospheric effects of Tambora's huge explosion and caldera collapse, 1816 was remembered in Europe and America as the "year without a summer"—a year when snow fell in New England in June.

The idea that a volcanic eruption could affect the world's climate surfaced again after the 1883 eruption of Krakatau. This time the visible atmospheric effects were unquestionably worldwide; they began within

149 Cinder cones along the Laki fissure in Iceland. This fissure erupted in 1783 with the greatest lava flood in recorded history. The gases released during this enormous eruption caused a blue haze or "dry fog" that reached Europe.

two weeks of the great explosive eruption and lasted for months. Strange colors and halos around the sun and moon were noted, and there were vivid sunrises and sunsets for months on end (Figure 150). A Ceylon (now Sri Lanka) newspaper for September 17, 1883, gave this account:

> The sun for the last three days rises in a splendid green when he is visible; about 10° above the horizon. As he advances he assumes a beautiful blue, and as he comes further on looks a brilliant blue, resembling burning sulfur . . . even at the zenith, the light is blue, varying from pale blue to a light blue later on, somewhat similar to moonlight . . . Then as he declines, the sun assumes the same changes, but vice versa.

The eruption appears to have lowered world temperatures for two to three years by as much as 0.5°C below normal (Figure 151). Some people have questioned these data, however, as most weather observations in the 1880s were made in Europe, which indeed was colder, but South

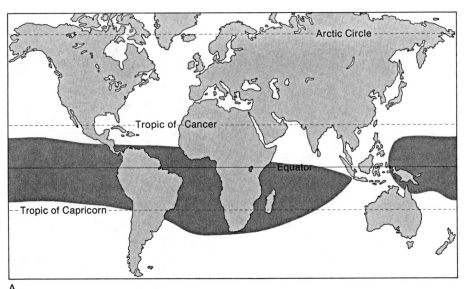

A

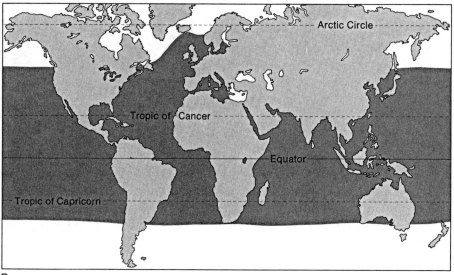

B

150 High-speed stratospheric winds averaging nearly 120 kilometers per hour carried the fine volcanic dust of the 1883 Krakatau eruption westward around the globe. By the end of November, the stratospheric haze covered over 70 percent of the Earth's surface, causing spectacular sunsets and strange optical phenomena. A. The approximate distribution of these atmospheric phenomena between August 26 and September 7, 1883. B. The approximate limits of the main atmospheric phenomena at the end of November 1883. (Adapted from C. J. Symons, ed., *The Eruption of Krakatoa*, Royal Society Report of the Krakatoa Committee, 1888.)

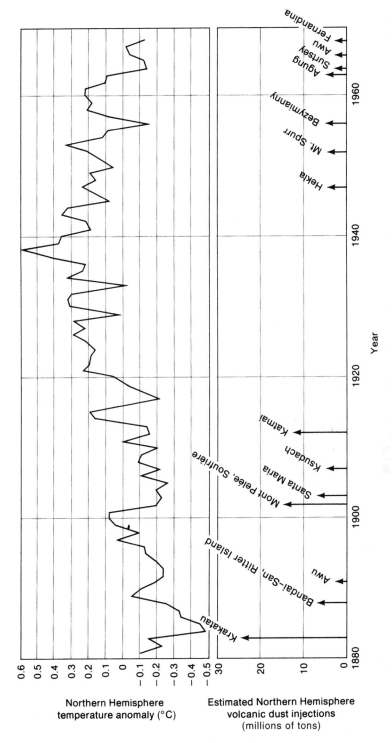

151 A graph of the average temperature variations in the Northern Hemisphere and the major volcanic eruptions that injected dust into the stratosphere. (Adapted from Robert Oliver, *Journal of Applied Meteorology* 15 [1976]: 934.)

America might have been warmer, and thus the true average world effects were not well known.

Harry Wexler, a former chief scientist of the U.S. Weather Bureau, was a strong proponent of the idea that volcanic eruptions affect climate. In 1952 he suggested that the warming trend of world temperatures since 1900 might be caused by the lack of major volcanic eruptions in the first half of the twentieth century. By contrast, he blamed the colder decades closing the nineteenth century on a series of major eruptions: Krakatau in Indonesia, 1883; Tarawera in New Zealand, 1886; Bandai-san in Japan, 1888; and Bogoslof in Alaska, 1890.

H. H. Lamb, a British climatologist, also has been a champion of the idea that volcanic activity affects climate. He compiled a detailed list of volcanic eruptions since A.D. 1500 and computed a "dust veil index" based on the apparent amount of volcanic debris scattered into the atmosphere. He concluded that there has been a definite relationship between world climatic trends and large volcanic eruptions.

In addition, J. P. Kennett and R. C. Thunell, working with cores from deep-sea drilling projects, concluded that the amount of volcanic ash in seafloor sediments increased about 2 million years ago and has stayed high since then. The last 2 million years coincide with the Pleistocene ice ages, and Kennett and Thunell concluded that the extra volcanism and extra cold are not just a coincidence. But even this evidence is not conclusive; some scientists point out that not enough deep-sea sediments older than 2 million years have been collected to make a valid comparison.

If volcanic gas and dust do alter weather and climate, the effect probably operates in the stratosphere (above 15 kilometers), where the layer of haze hovers for a long time, since there are no clouds and rain to wash it away quickly (Figure 152). Meteorologists have identified a long-lasting stratospheric aerosol layer at heights of 15 to 30 kilometers that seems to be composed of a thin haze of small particles or droplets, smaller than 1 micron in diameter. These particles are made up of various materials, including sea salt, silicate dust, and sulfuric acid. They probably originate from several sources—sea spray, dust storms, volcanic eruptions, forest fires, industrial smokestacks, and so on. The density of the aerosol layer changes over periods of months to years. It can increase suddenly with an injection of new aerosol from a volcanic eruption, but it takes several years to decrease to normal.

This layer of haze in the stratosphere intercepts the incoming sunlight, heating the stratosphere and cooling the lower atmosphere as well as the Earth's surface. Measurements by the National Oceanic and Atmospheric Administration's Mauna Loa Observatory show decreases in

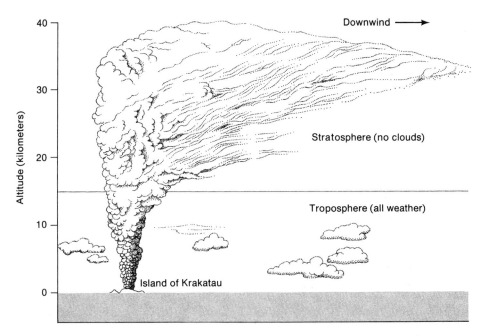

152 The Krakatau eruption cloud of 1883 was injected high into the stratosphere, where no rain clouds exist to wash the debris back to Earth. Once there, the fine dust stays in suspension for months to years. (Adapted from Richard Hazlett, used with permission.)

solar transmission through the atmosphere just after the 1963 eruption of Agung Volcano on the Indonesian island of Bali, the 1982 eruption of El Chichón Volcano in Mexico, and the 1991 eruption of Mount Pinatubo in the Philippines (Figure 153). These three volcanoes produced high eruption clouds that reached well into the stratosphere, and all three were located at low latitudes, so that the haze from their eruption clouds was able to reach both hemispheres. In addition, the El Chichón and Pinatubo eruptions released unusually large amounts of sulfur dioxide, which oxidizes to sulfuric acid aerosol in the atmosphere.

As can be seen in Figure 153, the May 18, 1980, eruption of Mount St. Helens had no apparent atmospheric effect in Hawaii. This is attributed to three factors: except for the very beginning of the eruption, the cloud barely reached the bottom of the stratosphere; the location of Mount St. Helens is far from the Earth's equator; and the emission of sulfur dioxide was relatively low for the size of the eruption.

El Chichón Volcano in southern Mexico, dormant since prehistoric times and thought to be extinct, blasted into action in late March 1982. In one climactic week, three major explosions lofted large ash and gas

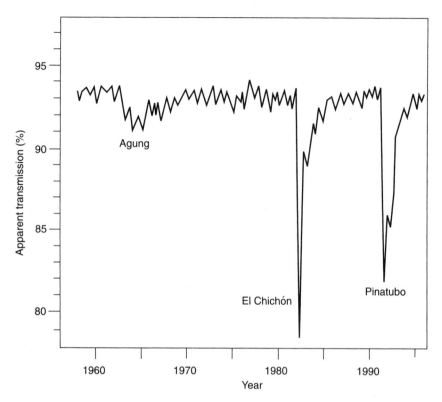

153 The percentage of direct solar radiation (apparent transmission) arriving at the Mauna Loa Observatory on clear days over several decades. Stratospheric haze from explosive volcanic eruptions of Agung Volcano, El Chichón Volcano, and Mount Pinatubo reduced direct solar radiation by scattering the incoming sunlight. The last two eruptions involved magma rich in sulfur, which produced haze containing large amounts of sulfuric acid aerosol. Sulfuric acid droplets in stratospheric haze are particularly effective in scattering solar radiation. The apparent transmission line is smoothed from monthly averages measured at 3415 meters elevation on the north flank of Mauna Loa Volcano, Hawaii. (Graph courtesy of the Mauna Loa Observatory, NOAA, CMDL.)

clouds to heights of 20 to 25 kilometers. Two to three thousand people in villages near El Chichón were killed by pyroclastic flows that accompanied the eruption. The Mount St. Helens eruption was larger in terms of the volume of rock involved, but El Chichón's atmospheric effects were far greater. Besides fine ash particles, El Chichón injected an enormous amount of sulfur dioxide into the stratosphere. This gas soon oxidized, picked up water vapor, and formed tiny droplets of sulfuric acid. The sulfuric acid aerosol and fine ash were carried westward by high-speed stratospheric winds, circling the Earth within a month.

Measured by LIDAR in Hawaii, the El Chichón stratospheric haze was 100 times more opaque than the Mount St. Helens cloud had been. (LIDAR is a technique of aiming a laser beam into the clear night sky and measuring the backscatter of laser light that is reflected from fine particles or aerosol droplets suspended in the atmosphere.) It took nearly a year for the El Chichón stratospheric cloud to attain its maximum effect over Europe, and it was not until 1985 that solar transmission measurements returned to pre-eruption levels.

The most convincing evidence of the connection between volcanoes and world climate came in June 1991, with the huge eruption of Mount Pinatubo in the Philippines (see Chapter 5). Satellite images showed that the climactic June 15 explosions produced a mushroom cloud of ash 400 kilometers wide and 34 kilometers high. The main explosive jetting of the ash eruption lasted for three hours, and the overall magnitude of the eruption was ten times larger than the 1980 eruption of Mount St. Helens. The enormous cloud of ash and gas that spewed from Pinatubo reached high into the stratosphere, above the reach of clouds and rainfall that could wash it out.

The entrapped gases that caused Pinatubo's explosion were rich in sulfur dioxide. Oxidized in the atmosphere to small aerosol droplets of sulfuric acid, this haze completed its circle around the Earth in three weeks and then slowly spread over the sphere during the next year. The aerosol cloud reached its maximum worldwide distribution by late 1992, reducing the amount of sunlight reaching the Earth's surface during 1992–1993. The aerosol slowly decreased to more normal levels by 1995. Average world temperatures dropped about 0.4°C during 1992–1993, bucking the trend of global warming during the past two decades.

The Earth has been in a warming trend since 1980. Most meteorologists blame this trend on the increasing amount of carbon dioxide in the atmosphere, caused by the burning of fossil fuels and the destruction of rain forests that consume carbon dioxide. Carbon dioxide in the atmosphere has a "greenhouse effect": it is transparent to light but more opaque to infrared radiation. Although the warming of the Earth by sunlight during the day is not affected, the cooling of the Earth at night by infrared radiation into space is impeded. The carbon dioxide acts like the windows in a greenhouse: light can come in, but heat cannot escape. The net effect is a global warming of the Earth's surface. Without this greenhouse effect, the cooling influence of the volcanic haze from El Chichón and Pinatubo might have been much more pronounced.

The question remains: To what extent do volcanic eruptions affect our weather and climate? Evidence shows that eruptions injecting large amounts of fine ash and, even more important, great quantities of sulfur dioxide gas into the stratosphere do significantly affect the solar radiation reaching the Earth's surface. Major explosive eruptions, particularly in equatorial regions where their ash and gas clouds may affect both hemispheres, appear to be capable of lowering the average global temperature by as much as 1°C. An extremely large event like the Yellowstone eruption 600,000 years ago, or a series of major explosive eruptions that prolong the stratospheric haze layer for decades or centuries, would have a much greater cooling effect.

Volcanic eruptions are only one factor among many that affect weather and climate; the normal interaction of the atmosphere, oceans, and land surface is extremely complex. Add to this the disruption caused by natural phenomena and by humans, and the intricacy of the problem becomes apparent. It is a joint task for geologists, meteorologists, and climatologists to understand and quantify these effects. Such understanding is of immense importance to a world that depends on a benign climate for its very existence.

Links

Aerosols and Climate
  http://www.ogp.noaa.gov/aboutogp/spotlight/aerosols/aerofigure1.htm
EOS Plume Studies
  http://eos.higp.hawaii.edu/ppages/self.html
Global Carbon Budget
  http://pubs.usgs.gov/fs/fs137-97/index.html
Global Climate Change
  http://www.exploratorium.edu/climate/index.html
Krakatau 1883 Eruption
  http://vulcan.wr.usgs.gov/Volcanoes/Indonesia/description_krakatau_1883_eruption.
  html
NASA EOS Volcanology
  http://eos.higp.hawaii.edu/
Sulfur Dioxide Aerosols
  http://volcanoes.usgs.gov/Hazards/What/VolGas/SO2Aerosols.html
TOMS Volcanic Emissions Group
  http://toms.umbc.edu/
Volcanoes & Climate Change
  http://earthobservatory.nasa.gov/Study/Volcano/
Volcanic Gases and Climate
  http://pubs.usgs.gov/of/of97-262/of97-262.html
Volcanoes and Weather
  http://vulcan.wr.usgs.gov/Glossary/VolcWeather/description_volcanoes_and_weather.
  html

Sources

American Geophysical Union. *Volcanism and Climate Change*. AGU Special Report. Washington, DC: American Geophysical Union, 1992.

Angell, J. K. "Comparison of Stratospheric Warming Following Agung, El Chichón, and Pinatubo Volcanic Eruptions." *Geophysical Research Letters* 20 (1993): 715–718.

Carey, S., and M. Bursik. "Volcanic Plumes." In H. Sigurdsson, ed., *Encyclopedia of Volcanoes*. San Diego, CA: Academic Press, 2000.

Defoor, T. F., E. Robinson, and S. Ryan. "Early LIDAR Observations of the June 1991 Pinatubo Eruption Plume at Mauna Loa Observatory, Hawaii." *Geophysical Research Letters* 19 (1992): 187–190.

Lamb, H. H. *Climate: Present, Past, and Future*. London: Methuen, 1972.

McCormick, M. P., L. W. Thomason, and C. R. Trepte. "Atmospheric Effects of the Mt. Pinatubo Eruption." *Nature* 373 (1995): 399–405.

Mills, M. J. "Volcanic Aerosol and Global Atmospheric Effects." In H. Sigurdsson, ed., *Encyclopedia of Volcanoes*. San Diego, CA: Academic Press, 2000.

Rampino, M., and S. Self. "Volcanism and Biotic Extinctions." In H. Sigurdsson, ed., *Encyclopedia of Volcanoes*. San Diego, CA: Academic Press, 2000.

Rampino, M., S. Self, and R. B. Stothers. "Volcanic Winters." *Annual Review of Earth and Planetary Sciences* 16 (1988): 73–99.

Robock, A., and J. Mao. "The Volcanic Signal in Surface Temperature Records." *Journal of Climate* 8 (1995): 1086–1103.

CHAPTER 18

Forecasting Volcanic Eruptions

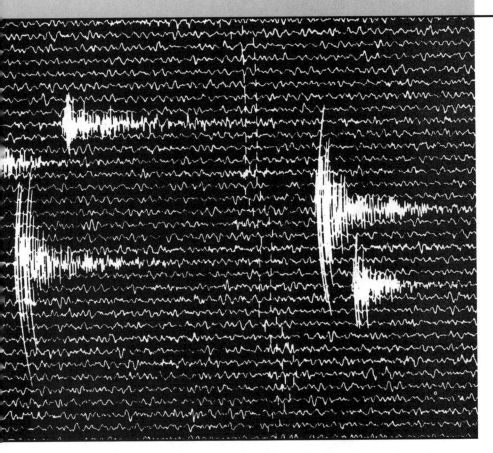

154 A seismogram of small earthquakes related to volcanic activity in Hawaii. (Photograph by the U.S. Geological Survey.)

Forecasting the timing, place, and character of volcanic eruptions is one of the major goals of volcanology. We prefer the word *forecast* to *prediction* because the science of weather forecasting has established that forecasts are probabilistic; that is, they are not precise, though the hope is that they are more accurate than statistical averages. For example, suppose that rainfall records in Hilo, Hawaii, show that on average it rains during six out of ten days throughout the year. If a weatherman forecasts a 60 percent chance of rain tomorrow in Hilo, he is not making a very adventurous statement. But if the wind patterns and satellite photos indicate that the chances of rain tomorrow are greater or less than average, then the forecast may say a 90 percent chance of rain or a 10 percent chance of rain. Such a forecast would be valid based on information other than historical statistics; its worth could be evaluated in hindsight by comparison with random guesses. Weather forecasters avoid the word *prediction* because it sounds so precise and specific. Even though current weather forecasts are far from exact, they are extremely valuable. They do much better than random guesses, and their batting average is improving.

The current goal in forecasting volcanic eruptions is to provide the best forecasts possible, based on the geologic history of the volcano under study as well as on the day-to-day vital signs of the volcano, including earthquakes, surface deformation, temperature, and gas emissions. The statistics of the volcano's past eruptions are of great importance, both as a basis for determining its average probability of eruption and as a means of deciphering some pattern in its eruptive habits. At least three such patterns have been recognized, even though the periods of repose between eruptions may vary greatly.

One pattern is completely random. That is, no matter how long the repose period has been, the average chance for an eruption next month remains the same. This is like cutting cards to get an ace; no matter how many times you fail, the chance in the next cut is exactly the same: 4 out of 52. Mauna Loa Volcano in Hawaii appears to operate in this random manner. For most of its known eruptive history its average repose time has been about 4 years. Thus, no matter how long or short the repose between eruptions, the chance for a new eruption during the next

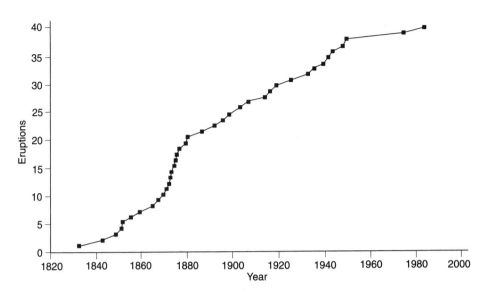

155 The cumulative number of historical eruptions of Mauna Loa Volcano in Hawaii plotted against their onset dates. The average time between eruptions is four years. However, the frequent eruptions between 1870 and 1880 and the long interval after the eruption of 1950 suggest distinct clusters of nonrandom recurrence times.

month would remain the same—2 percent (Figure 155). However, the three very long repose periods since the 1950 eruption challenge this overall random interpretation.

Hekla Volcano in Iceland shows quite a different pattern of time intervals between eruptions. At Hekla the average probability of an eruption increases with time. This is like cutting for aces and discarding the cut card each time you fail, thereby increasing the chance of getting an ace with each new cut.

Just the opposite holds for volcanoes like Kilauea in Hawaii, where groups of eruptions cluster together in time (Figure 156). In this situation, the probability of an eruption decreases with time. There is no easy analogy for this in card cutting.

Each of these patterns of eruptive habits is important as a basis for forecasting future activity. Unfortunately, for the statistics to have any meaning, the number of eruptions must exceed 10 or 20, and only a small fraction of the world's volcanoes are active enough, or have been studied long enough, to establish their eruptive patterns. It is possible, however, to estimate how often volcanic eruptions of differing sizes will occur worldwide. Of course, the number of small eruptions greatly exceeds that of large ones, but this number can be quantified. Table 8,

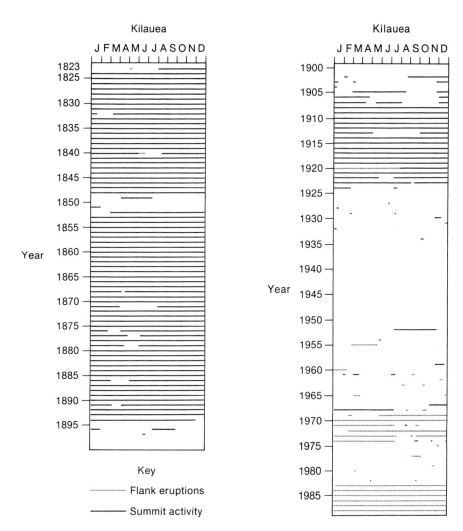

156 Eruptions of Kilauea Volcano in Hawaii since 1823 show a pattern of long periods of continuing eruptions or groups of eruptions interspersed with long periods of repose. The periods from 1823 to 1894 and from 1906 to 1924 were characterized by nearly continuous lava lake activity in the summit crater. The pattern that began in 1983 continues to the time of this writing (March 2005). (After Gordon Macdonald and Douglass Hubbard, *Volcanoes of the National Park of Hawaii*, Hawaii Natural History Association, 1982.)

based on statistics compiled by the Smithsonian Institution and the U.S. Geological Survey (USGS), indicates the number of explosive eruptions of a particular magnitude or larger that have taken place during the past 200 years, plus the extremely large explosive eruptions that apparently have occurred during the past million years. Extrapolating this table into

TABLE 8   Number versus size of explosive eruptions

VEI[1]	Volume	Number	Adjective	Example
1	10^4–10^6 m^3	100 per year	Small	Stromboli 1996
2	10^6–10^7 m^3	15 per year	Moderate	Unzen 1991
3	10^7–10^8 m^3	2–3 per year	Moderate–large	Ruiz 1985
4	10^8–10^9 m^3	1 per 2 years	Large	Rabaul 1994
5	1–10 km^3	1 per 10 years	Very large	Mount St. Helens 1980
6	10–100 km^3	1 per 40 years	Huge	Pinatubo 1991
7	100–1000 km^3	1 per 200 years	Colossal	Tambora 1815
8	1000–10,000 km^3	1 per 50,000 years	Humongous	Yellowstone 630,000 B.C.
9	>10,000 km^3	Not found	Fortunately	None

[1]Volcanic Explosivity Index (see Table 3 in Chapter 9)

the future gives a rough estimate of how often eruptions of various sizes will take place in the years and centuries to come.

Dormant volcanoes are by no means dead, and by studying them with sensitive instruments it is possible to monitor their vital signs through periods of repose and awakening. One of the most important of these signs is the number of earthquakes and their locations.

Both earthquakes and volcanoes occur along plate margins, where many of the earthquakes are considered to be related to the slow slipping and breaking of the plates' moving edges. Some earthquakes, however, appear to be more directly related to volcanic processes. In Hawaii and other areas of hot-spot volcanoes, earthquakes accompany volcanism, even though the plate margins may be thousands of miles away (Figures 154 and 157). In fact, Hawaii's Kilauea Volcano is one of the most seismically active areas on Earth. Ten or more microearthquakes occur beneath the volcano on an average day, and there is an earthquake large enough to be felt about once a week.

During the immediate prelude to the eruption of a volcano, when new magma conduits are opening underground, hundreds to thousands of microearthquakes shake seismometers, with several felt shocks generally accompanying these earthquake swarms. At subduction volcanoes like Mount St. Helens, the number of microearthquakes during dormant periods is usually only a few per month, but in the swarms preceding an eruption the count of small quakes generally increases to several hundred per day.

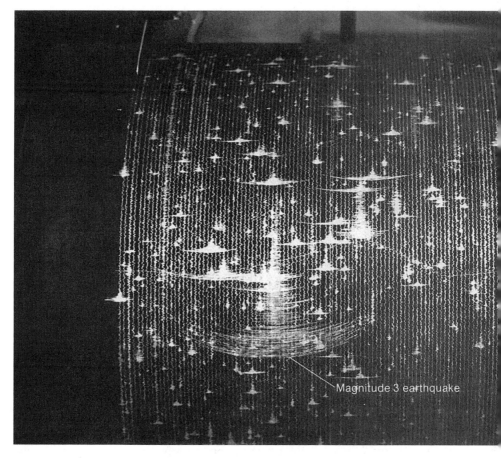

157 A seismograph recording earthquakes at the Hawaiian Volcano Observatory. The drum rotates once every 15 minutes, slowly enough that an entire day can fit on one record sheet. Most of the microearthquakes on this record are too small to be felt. However, the larger quake indicated, which was about magnitude 3, was large enough to be felt in a local area close to its source. (Photograph by the U.S. Geological Survey.)

There is probably more than one cause for these volcanic earthquakes, including increasing topographic load, underground temperature changes, moving magma, and gas explosions. The slipping and cracking of rocks underground to adjust to the growing weight of a huge volcano is only indirectly related to volcanic activity, but the other causes, particularly the movement of magma and the formation of cracks through which it can move, are closely connected to active volcanic processes.

Studies by Bernard Chouet of USGS indicate that swarms of long-period (LP) earthquakes occurring at shallow depths beneath a volcano seem to be useful indicators of an impending eruption. These precursors may occur over periods as short as days to weeks before an eruption. Swarms of such LP earthquakes have clearly preceded recent eruptions at Redoubt Volcano in Alaska, Mount Pinatubo in the Philippines, and Galeras Volcano in Colombia.

The number or size of volcanic earthquakes, particularly those related to the conduits through which magma erupts to the surface, usually increases before an eruption. This relationship is not infallible, however. In a study of 71 earthquake swarms and volcanic eruptions, 58 percent of volcanoes showed an increase in earthquake activity before an eruption, 38 percent showed an increase without an eruption, and 4 percent showed no apparent increase in earthquake activity before an eruption. This study by David Harlow, who was then a student at Dartmouth College, dates back to 1970. Since then, there have been many more earthquake swarms and eruptions, and a 1996 study by John Benoit and Stephen McNutt at the University of Alaska shows remarkably similar results. Their survey of 327 earthquake swarms beneath potentially active volcanoes between 1979 and 1989 found that 191 of the swarms were followed by eruptions (58 percent) and 136 (42 percent) were not. The swarms not followed by eruptions had a shorter average duration (3.5 days) than did the precursory swarms (8 days).

Because the background count of microearthquakes in volcanic areas is highly variable, only a large change in their number—by as much as a factor of 100—seems to be significant. Russian scientists in Kamchatka now believe that large increases in the total energy released by volcanic earthquakes are more important than increases in their number. Barry Voight, at Pennsylvania State University, argues persuasively that it is the acceleration of the earthquake count or energy release that is the key figure. The time between the beginning of an earthquake swarm and the actual eruption varies from months to hours, but even so, the number, size, and location of earthquakes on active volcanoes provide an important index of forthcoming activity.

Volcanic tremor is a unique kind of seismic activity associated with volcanoes. It consists of more or less continuous ground vibration with a frequency of 0.5 to 10 cycles per second—a very low hum detectable by seismographs (Figure 158). Its source is not clear; various studies relate it to the formation of gas bubbles or the irregular flow of magma, which creates vibrations like the noise of water running in poorly designed pipes. Whatever its source, it is nearly always present during a volcanic eruption and often begins before the actual surface outbreak. Not all periods of volcanic tremor are followed by an eruption, but in Hawaii, high-amplitude volcanic tremor is often the best indication that an eruption

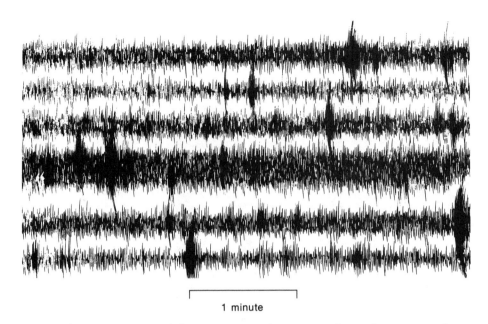

1 minute

158 Volcanic tremor recorded on a seismograph in Hawaii during the eruption of Kilauea in 1977. This low-frequency vibration of the ground at about 3 to 5 cycles per second is related to the movement of magma through underground conduits. The tremor starts before the breakout of an eruption and is a sign that an eruption is imminent. (Photograph by James Griggs, U.S. Geological Survey.)

has begun or is about to begin within a few minutes or hours. A "tremor alarm" ringing in their houses alerts the observatory staff to check their instruments.

Slight changes in the angle of slope or distance between survey points on the summits and flanks of active volcanoes provide another important means of diagnosing internal changes taking place. Several techniques are used to detect such changes, including conventional level surveys and the determination of distances with reflected laser beams. Tiltmeters, which can detect changes in slope smaller than 1 part per million, are also used. An angular change in slope of 1 part per million (1 microradian) is equivalent to lifting the end of a rigid beam 1 kilometer long by only 1 millimeter. A technique called GPS (Global Positioning System) uses satellites and portable receivers to determine the relative positions of ground-surface survey stations to accuracies of a few millimeters (Figure 159). Radio code and time signals broadcast by the satellites arrive at the survey stations in a unique pattern. A comparison by computer processing of the different patterns recorded at two survey stations deter-

159 A GPS (Global Positioning System) antenna is set up over a survey benchmark to receive radio time and code signals from passing satellites. Elevations and distances between two or more receivers operating simultaneously can be determined to accuracies of better than 1 part per million.

mines their relative locations. GPS receivers that monitor changes in position almost continuously have recently been installed at Augustine Volcano in Alaska and at Kilauea Volcano in Hawaii.

Both elevation and distance changes can be detected by GPS measurements. The receivers can be many kilometers apart and need to "see" the satellites, but not one another. Reasonably priced GPS instruments are now available, and important observations can be made in a few hours by a small group of surveyors. GPS surveys in Hawaii have revealed that the southeast flanks of both Kilauea and Mauna Loa are slowly moving a few centimeters per year toward the southeast (Figure 160). In 2001–

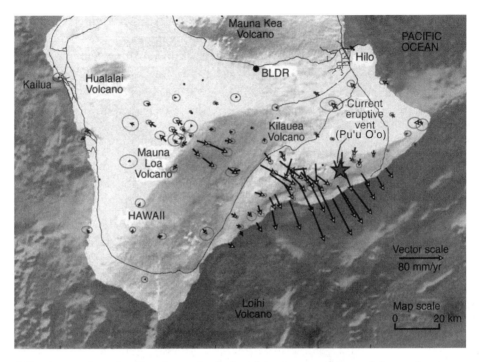

160 Arrows show the direction and velocity of the slow movement of the southeast flank of Kilauea Volcano toward the sea. This movement has been measured using Global Positioning System (GPS) receivers from 1990 through 1995. The ellipses show the uncertainty of the measurements. The point labeled BLDR is the reference station. (Data and map from Mike Lisowski, Hawaiian Volcano Observatory, U.S. Geological Survey.)

2002, additional continuous GPS receivers were installed on Kilauea and Mauna Loa that provide nearly real-time data on these volcanoes.

The latest technique used to "see" ground-surface deformation on volcanoes is called *satellite radar interferometry* (InSAR). Digital radar images of the topography, taken at different times, are subtracted from one another. If uplift or subsidence has occurred, the changes appear as interference rings or bands of iridescent colors, like the colored rings or bands seen in the reflected light from a thin oil slick on water. If there has been no deformation, no interference bands will appear. Although still in an experimental phase, this technique seems to have great promise. Satellite radar surveys at periodic intervals could monitor many potentially dangerous volcanoes and also provide basic research data on their internal structure and dynamics.

Repeated at frequent intervals, any of these surveying techniques may reveal the tiny deformations of the volcano's surface caused by changes in magmatic pressure or volume inside the volcano. *Deformation moni-*

toring, the collective name for all these surveying techniques, has become the second most important means of forecasting volcanic eruptions, preceded only by the monitoring of earthquakes.

The pattern of slow inflation of Kilauea Volcano in the months before an eruption, and its sudden collapse over a period of hours or days during an eruption (described in Chapter 7), has been revealed by deformation measurements over the past 40 years (Figure 161). The slow, more or less continuous addition of magma to the shallow chamber beneath Kilauea's summit causes the summit to swell upward as much as 2 meters over a diameter of 10 kilometers. This gentle bulge causes outward tilts and increasing distances between survey stations, similar to the changing relationships of spots on the surface of an inflating balloon. Intermittent eruptions of Kilauea between 1955 and 1986 removed molten lava from the magma chamber faster than it was replenished, causing summit deflation, inward tilt, and contracting distances between survey stations. Since 1986, Kilauea has been erupting almost continuously, and magma is being replenished nearly as fast as it is erupted. During the intermittent eruptions, deformation measurements provided a kind of magma barometer that in turn could be used to help forecast eruptions. If there were an exact level of reinflation that triggered a new eruption, the question of when an eruption will occur would be solved.

But nature is not that simple. On Kilauea, which has been tracked through more deformation cycles than any other single volcano, the degree of tilt at the Hawaiian Volcano Observatory just before an eruption has always been relatively high, but has not shown any specific critical angle. The reason for this may be that an increase in either pressure or volume in the shallow magma chamber beneath Kilauea can cause uplift and outward tilt of the summit area. Slow increases in volume may not cause equal increases in pressure, and it is increasing pressure that causes the hard rocks surrounding the magma chamber to split apart. Magma then moves into these newly formed underground cracks, and the summit subsides rapidly. The magma may halt underground in these new fractures as a shallow intrusion, or it may break to the surface in an eruption.

Deformation measurements on several potentially explosive volcanoes on subducting plate margins are now being made, but it is not yet clear whether they will be as useful as those on basaltic shield volcanoes. One problem with explosive subduction volcanoes is their long repose time between eruptions. It could take several hundred years to learn as much about deformation at Vesuvius as has already been learned at Kilauea in the last 40 years. Nevertheless, monitoring of both the earthquake swarm and the large surface bulge that took place for nearly two

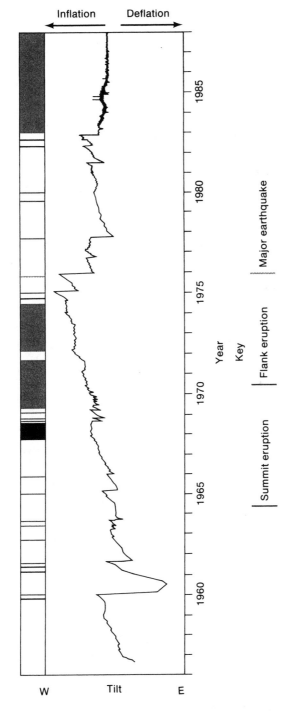

161 The 40-year tilt record of Kilauea Volcano in Hawaii shows the ups and downs related to eruptions. For months to years before an eruption, the summit of Kilauea inflates. When it reaches a level that exceeds the strength of the rocks surrounding the shallow magma chamber, the volcano erupts. Some eruptions rapidly remove large volumes of magma from the summit chamber, and the volcano deflates over a period of days. A high level of inflation indicates a higher probability of eruption, but since the strength of the rocks surrounding the magma chamber varies from one eruption to another, no exact level of tilt can be used to predict the next eruption. (Data from the U.S. Geological Survey.)

months before the major eruption of Mount St. Helens in 1980 provided important insights into what was going on beneath the volcano during that period (see Chapter 4).

Changes in the Earth's magnetic and electrical fields near volcanoes that relate to the state of volcanic activity have been observed in Japan, New Zealand, Kamchatka, and Hawaii. Although the techniques used for these observations are not as well established as seismic and deformation measurements, they look promising.

Temperature and compositional changes at steam vents and warm springs on volcanoes would seem to be an obvious index for volcano forecasting, but rainfall and changes in groundwater circulation often cause large fluctuations in temperature not related to volcanic activity. In one case, however, a 12°C rise in the temperature of the crater lake at Taal Volcano in the Philippines clearly signaled its 1965 eruption (Figure 162). Geochemical changes in the volume and composition of volcanic gases also are useful indicators of hidden changes beneath active volcanoes. New steam vents formed at Askja Volcano in Iceland two weeks before its eruption in 1961. Scientists in Japan, Kamchatka, and the United States all have reported an increase in sulfur gases relative to chloride gases at volcanic steam vents in the years or days before some eruptions. Changes in the percentages of hydrogen, helium, and radon in volcanic gases are also being studied as possible signals of changing volcanic activity.

No single technique appears to be the master key to forecasting volcanic eruptions. Each volcano is unique, and the case history of one cannot always be used to diagnose the symptoms of another. Even so, useful, though not precise, forecasting is currently being practiced on volcanoes in Japan, Indonesia, Iceland, the Philippines, Kamchatka, Hawaii, and Alaska. The success in forecasting the 1975 and 1984 eruptions of Mauna Loa Volcano in Hawaii illustrates the present state of the art.

Mauna Loa Volcano erupted on July 5 and 6, 1975, after 25 years of repose—its longest sleep since records began in the mid-nineteenth century. Thirty million cubic meters of lava poured out in that brief but intense summit eruption. The monitoring of earthquakes and deformation at Mauna Loa was increased during and after the eruption, and scientists waited with interest to find out what these observations would reveal while Mauna Loa rested and prepared itself for the next eruption. It was a 10-year wait, and Figure 163 sums up the data.

Small earthquakes of shallow and intermediate depth showed an increase for one to two years before both the 1975 and 1984 eruptions, but the number of earthquakes deeper than 13 kilometers showed no change.

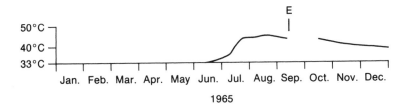

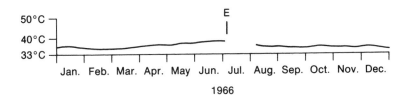

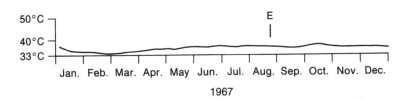

162 Changes in the temperature of the crater lake on Taal Volcano in the Philippines sometimes warn of forthcoming eruptions. In 1965 the temperature began to rise well above its background level of 33°C in July; the volcano erupted in September. The 1966 eruption was preceded by a much less obvious temperature rise, and the 1967 eruption could not have been forecast on the basis of temperature alone. (Data from A. Alcaraz, Philippine Commission on Volcanology.)

The sharp increase in intermediate-depth earthquakes in late 1983 was caused by numerous aftershocks that followed a magnitude 6.6 earthquake that occurred beneath the east flank of Mauna Loa in November of that year.

Deformation measurements of the summit showed a 75-millimeter increase in the length of a 3-kilometer-long survey line across the caldera in the year before the 1975 eruption, as well as a similar increase before the 1984 eruption. At the onset of both eruptions, a sudden widening of the summit caldera by about 500 millimeters was observed, caused by the opening of magma-filled cracks that split the caldera and fed the eruptions.

Following the start of the eruption in 1984, the width of the caldera contracted by 300 millimeters as the Northeast Rift Zone vents at 2900

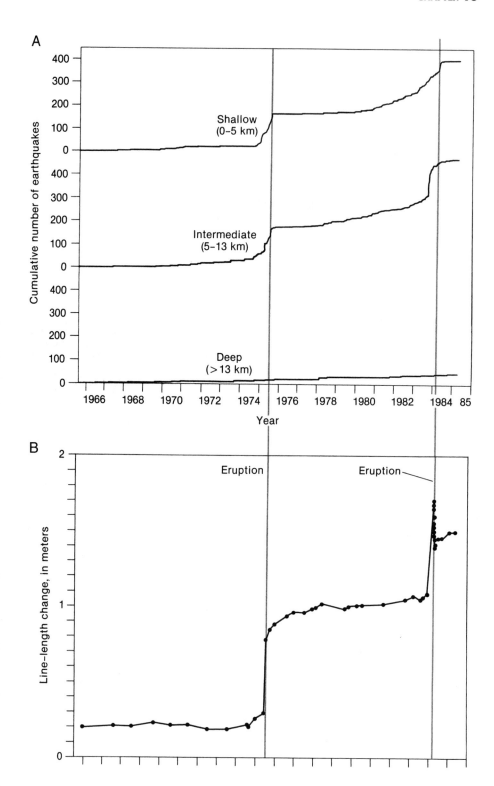

163 Earthquakes and deformation are precursors to eruptions of Mauna Loa in Hawaii. A. Shallow and intermediate-depth quakes beneath the volcano, most of them too small to be felt, increase for one to two years before an eruption. B. The 3-kilometer-wide caldera widens mainly during eruptions (each dot represents a separate survey). (After J. P. Lockwood et al., U.S. Geological Survey Professional Paper 1350, 1987, pp. 548–549.)

meters drained part of the summit magma reservoir and caused the summit to subside and contract. After the three-week-long 1984 eruption of 220 million cubic meters of lava (Figure 164), the summit began to reinflate and widen again. By 1997 the summit inflation and stretching had recovered about 50 percent of the subsidence and contraction that accompanied the flank eruption. Mauna Loa's long repose periods since

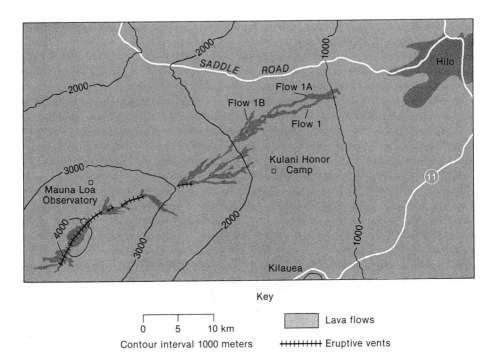

Key

0   5   10 km

Contour interval 1000 meters

Lava flows

++++++++ Eruptive vents

164 Fissure vents and lava flows of the 1984 Mauna Loa eruption. After the initial outbreak at the summit, eruptive fissures opened progressively down the northeast rift. The fractures at the 2900-meter elevation opened 15 hours after the eruption began and remained the principal vents of the 3-week-long eruption. Flow 1 reached its maximum length in 5 days and then diverted itself upstream into flow 1A. Flow 1B was formed by similar channel clogging of flow 1A on April 5, and the city of Hilo was no longer threatened by the eruption. (After J. P. Lockwood et al., U.S. Geological Survey Professional Paper 1350, 1987, p. 561.)

165 The great Tolbachik eruption in Kamchatka in 1975 was accurately forecast on the basis of the increasing energy of local earthquakes. The total volume of lava and pyroclastic material erupted was nearly 2 cubic kilometers, and the ash cloud reached 14 kilometers high. (Photograph by N. P. Smelov, Institute of Volcanology, Kamchatka.)

1950 suggest that the statistical probability of another eruption occurring in the next month has dropped to as low as 0.5 percent (1 chance in 200). The lack of any increase in shallow earthquakes beneath the summit of Mauna Loa in 2003 supports this forecast. Deeper earthquakes, however, increased in August 2004 and suggest that Mauna Loa may be awakening from its long repose.

The precursors of the 1975 and 1984 Mauna Loa eruptions were on two time scales: the earthquake increases that began about a year before the eruptions and the volcanic tremor that preceded the outbreaks by only one to two hours. Some precursors a few days or weeks before a Mauna Loa eruption would be most useful to discover.

The Russians may be doing better. The huge eruption of Tolbachik Volcano in Kamchatka in 1975 was preceded by a major earthquake swarm. Volcanologist P. I. Tokarev's published forecast that an eruption was imminent in the Tolbachik region during the next week allowed Russian television crews to be on hand for the birth of a new volcano on July 6, just two days after the forecast was issued (Figure 165).

The classic success story in volcano forecasting is the evacuation of the area around Mount Pinatubo in the Philippines just before its catastrophic eruption on June 15, 1991. That forecast, based on increasing seismicity and a prehistoric record of large explosive eruptions, saved tens of thousands of lives (see Chapter 5). Since 1962 our studies on active volcanoes have been directed toward forecasting. It is very rewarding to see the great progress made in this area of volcanology in the past 40 years.

Links

Alaska Volcano Observatory
http://www.avo.alaska.edu/avo4/index.htm

Cascades Volcano Observatory
http://vulcan.wr.usgs.gov/

Forecasting Eruptions
http://www.swvrc.org/forecast.htm

Forecasting Yellowstone
http://exodus2006.com/yellow.htm

Hawaiian Volcano Observatory
http://hvo.wr.usgs.gov/

Long Valley California Volcano Observatory
http://lvo.wr.usgs.gov/

Predicting Eruptions
http://vulcan.wr.usgs.gov/Vhp/C1073/challenge_predicting.html

Sensing Remote Volcanoes
http://earthobservatory.nasa.gov/Study/monvoc/

Tracking Yellowstone's Volcanic System
http://pubs.usgs.gov/fs/fs100-03/

Volcano Images, NASA
http://asterweb.jpl.nasa.gov/gallery/?catid=10

Vulnerable to Volcanoes
http://whyfiles.org/110fieldwork/2.html

Yellowstone Seismicity, University of Utah
http://www.seis.utah.edu/ynpreg.shtml

Yellowstone Volcano Observatory
http://volcanoes.usgs.gov/yvo/

Sources

Benoit, J. P., and S. R. McNutt. "Global Volcanic Earthquake Swarm Database and Preliminary Analysis of Volcano Earthquake Swarm Duration." *Annali di geofisica* 39 (1996): 221–229.

Chouet, Bernard. "Long-Period Volcano Seismicity: Its Source and Use in Eruption Forecasting." *Nature* 380 (1996): 309–316.

Decker, R. W. "Forecasting Volcanic Eruptions." *Annual Reviews of Earth and Planetary Sciences* 14 (1986): 267–291.

Decker, R. W. "How Often Does a Minoan Eruption Occur?" In *Thera and the Aegean World III*, Vol. 2, *Earth Sciences*. London: Thera Foundation, 1990.

McGuire, B., C. R. J. Kilburn, and J. Murry. *Monitoring Active Volcanoes*. London: UCL Press, 1995.

McNutt, S. R. "Seismic Monitoring." In H. Sigurdsson, ed., *Encyclopedia of Volcanoes*. San Diego, CA: Academic Press, 2000.

McNutt, S. R. "Volcano Seismology and Monitoring." In W. Lee et al., eds., *International Handbook of Earthquake and Engineering Seismology*. Boston: Academic Press, 2002.

Murray, J. B., H. Rymer, and C. A. Locke. "Ground Deformation, Gravity, and Magnetics." In H. Sigurdsson, ed., *Encyclopedia of Volcanoes*. San Diego, CA: Academic Press, 2000.

Rhodes, J. M., and J. P. Lockwood, eds. *Mauna Loa Revealed*. Geophysical Monograph 92. Washington, DC: American Geophysical Union, 1995.

Sabins, F. F. *Remote Sensing*. 3rd ed. New York: W. H. Freeman, 1997.

Smith, R. B., and L. Siegel. *Windows into the Earth: The Geologic Story of Yellowstone and Grand Teton National Parks*. New York: Oxford University Press, 2000.

Swanson, D. A., T. J. Casadevall, Daniel Dzurisin, R. T. Holcomb, C. G. Newhall, S. D. Malone, and C. S. Weaver. "Forecasts and Predictions of Eruptive Activity at Mount St. Helens, USA—1975–1984." *Journal of Geodynamics* 3 (1985): 397–423.

Voight, B. "A Method for Prediction of Volcanic Eruptions." *Nature* 332 (1988): 125–130.

CHAPTER 19

Reducing Volcanic Risk

166 Katia and Maurice Krafft, French volcanologists, at the Kupaianaha Lava Lake in Hawaii. They studied and photographed volcanoes all over the world, witnessing and documenting more eruptions than any other investigators. Their popular books, lectures, and videotapes greatly increased public awareness of the nature and dangers of volcanic eruptions. Ironically, along with American volcanologist Harry Glicken, they were killed by a pyroclastic flow from an eruption of Unzen Volcano in Japan in 1991.

The effort to forecast volcanic eruptions is not just an exercise in scientific curiosity. Its goal is to be able to reduce volcanic risk in an increasingly overpopulated world (Figure 167). With each decade, the number of people living close to—or on—dangerous volcanoes increases dramatically. Some are drawn by the wild, scenic beauty, others by the rich volcanic soils, and still others are pushed upslope by the population pressure on crowded islands.

The threat of a volcanic eruption in a densely populated area calls for a rapid response in many forms, including continuous monitoring of the awakening volcano's vital signs; assessment of potential hazards by studying its eruptive history and mapping the deposits around it; and—one of the most important components—good communication with local officials, the public, and news organizations. Residents generally wish to downplay the dangers; newspapers often exaggerate them, and scientists are squeezed in between.

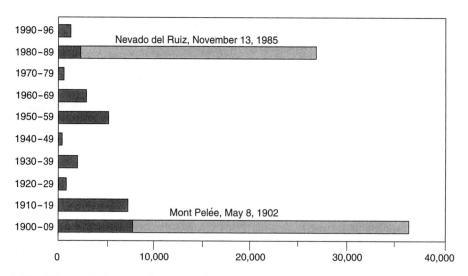

167 Volcano fatalities in the twentieth century through 1996. Most of the deaths occurred in just two events (indicated in light gray): the pyroclastic flow from Mont Pelée on the Caribbean island of Martinique and the mudflow from Nevado del Ruiz in Colombia. (After Robert Tilling, U.S. Geological Survey.)

In dealing with nature, especially in a form as capricious as a volcano, things seldom go as planned. One instance in which forecasting, contingency planning, and action all came together to achieve a remarkable success was the 1991 eruption of Mount Pinatubo in the Philippines (see Chapter 5). In contrast, the 1985 eruption of Nevado del Ruiz Volcano in Colombia, which killed 23,000 people, is an example of how badly things can go wrong (see Chapter 10). Most of the victims were residents of the town of Armero, at the foot of Ruiz. Geologists had warned them that because of the ice cap on the high summit and a lethal history of lahars, any eruption of molten or hot fragmental rock onto the summit ice could bring disaster. But no one in Armero really knew what a lahar or mudflow was; when the volcano erupted, only a few people evacuated the town. There was a two-hour period between the main summit eruption and the time the lahar swept over Armero, but the summit of Ruiz was not visible from Armero, and no one could see the approaching catastrophe. In hindsight, the tragedy might have been prevented. In reality, it was not.

This seems to have been a case of a massive failure of communication. Authorities knew of the critical danger from lahars, but failed to give an unambiguous warning to evacuate. The canyon above the village was not equipped with a flood-warning system—used commonly in Japan—which could have given ample warning. The use of a hazards video, which proved to be so effective in educating the people who were at risk from Pinatubo, could have made a life-or-death difference at Ruiz.

The eruptions of Mount St. Helens (described in Chapter 4) illustrate the strengths and weaknesses of volcano forecasting. Based on that volcano's geologic history, a warning was published in 1978 that destructive eruptions might occur within a few decades. After small eruptions began in March 1980, a strong, ongoing earthquake swarm and fast-growing bulge indicated to geologists monitoring Mount St. Helens that a sizable shallow intrusion of magma was being injected beneath the north flank of the mountain.

The probability that the magma would reach the surface in a significant eruption was considered quite high, but the timing and scale of the great May 18 eruption were not anticipated. Even in hindsight, no specific precursors or precedents can be detected that might have called for a complete evacuation of the area that was later devastated. Nevertheless, since the danger posed by the shallow intrusion of magma was recognized, access to the area was severely restricted, and many hundreds of lives were saved.

After the great eruption, there were five smaller explosions during 1980, and the lava dome in the crater grew rapidly during several peri-

ods between 1980 and 1987. Beginning with the July 1980 explosion, all
the eruptions and dome-building events at Mount St. Helens have been
forecast accurately from increases in seismicity and rates of deformation.
These precursors, which have occurred days to hours before each eruptive event, have given sufficient warning to allow evacuation of the hazardous area.

Most infamous volcanoes owe their notoriety to the deaths and destruction they have caused. Those eruptions that were forecast—thereby
saving thousands of lives—are less well known. One such example occurred in Indonesia in 1983. Colo Volcano, on the island of Una Una,
was shaken by an earthquake swarm that began on July 14, with small
explosive eruptions starting on July 18. On the basis of the past behavior
of Colo and similar volcanoes, geologists from the Volcanological Survey of Indonesia recommended to local government officials that the island be evacuated. The officials agreed, and all 7000 inhabitants were
removed by boat. On July 23 the climactic eruption swept the island with
hot pyroclastic flows. Most of the livestock and coconut plantations were
destroyed. It has taken years to rebuild the island's economy, but the people survived.

In general, volcanologists and public officials have an easier task getting public cooperation in a crisis in areas where the people have a mem-
ory of volcanic activity, and especially where there has been education
in preparedness (Figure 168). This was the case at Rabaul, on the island
of New Britain in Papua New Guinea, in 1994. The town of Rabaul is
situated on the shore of a bay that is actually a collapsed caldera. Its
location—on the Ring of Fire—is a clue that eruptions there have been
explosive. Other obvious clues to the island's volcanic origins are the
many volcanic cones that dot the caldera's rim. Two prominent cones,
Tavurvur and Vulcan, stand across the bay from each other like sentinels
guarding the entrance to the inner harbor (Figure 169).

Geologists believe that the caldera has collapsed several times in the
last 10,000 years. They have found charcoal fragments in pyroclastic deposits surrounding Rabaul Caldera with radiocarbon dates of about A.D.
500–600, representing its last huge collapse. Since that great eruption,
many smaller ones have built the cones scattered along the caldera rim.

In 1937 an eruption began in the shallow nearshore waters of the
caldera and built a 230-meter-high cone, Vulcan, most of it during the
first 12 hours. Hot pyroclastic flows from Vulcan killed more than 500
people. Tavurvur, 5 kilometers across the entrance of the bay, also erupted
during the brief 1937 event. Rabaul Town escaped the deadly pyroclastic flows, but was showered with 7 to 15 centimeters of volcanic ash.

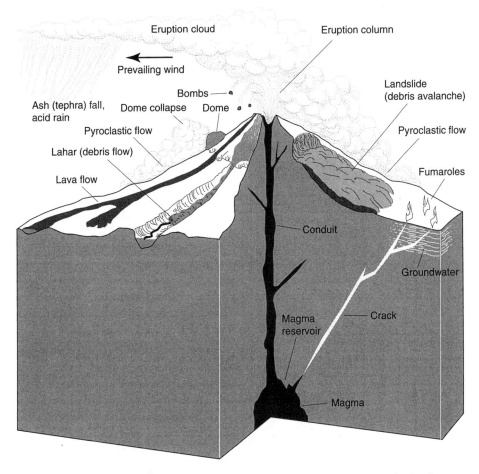

168   A sketch of the hazards posed by a typical stratovolcano. Volcanic risk, the danger from the hazards to living things and property, can be reduced by public awareness of volcanic processes and products. For example, living upwind from a volcano is safer than living downwind, and living in a valley near a volcano is more dangerous than living on a hillside. (From Bobbie Myers and Steven Brantley, U.S. Geological Survey.)

   The Rabaul Volcano Observatory (RVO), which was established before World War II but was closed during the Japanese occupation of Rabaul from 1942 to 1945, opened again in 1950. In 1973, scientists there began making some startling observations: the caldera floor was slowly bulging upward. Over the next decade they measured more than 1 meter of uplift. Small to moderate earthquakes—more than 90,000—and another half-meter of uplift were recorded between 1983 and 1984 (Figure 170). Since the geologists knew the violent history of the caldera,

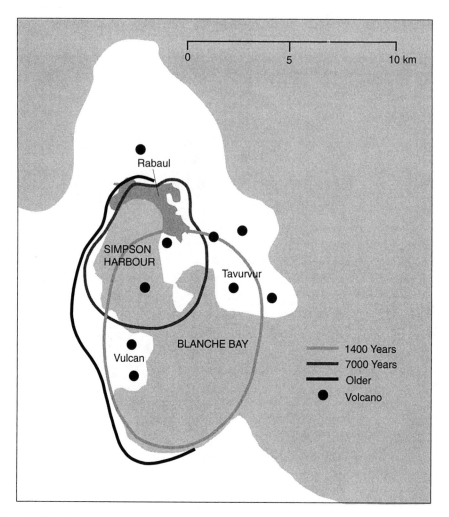

169 The sheltered harbor at Rabaul in Papua New Guinea is the result of multiple caldera collapses, the latest about 1400 years ago. Along with the benefits of a good harbor comes the danger of volcanic eruptions; the most recent occurred from both Vulcan and Tavurvur volcanoes in 1994. (After Russell Blong and Chris McKee.)

this unrest prompted disaster plans and practice evacuations and attracted volcanologists from all over the world. Then the earthquakes dropped from 10,000 back to tens or hundreds per month, and the uplift stopped. Fortunately, though, the RVO kept up its monitoring, practice evacuations, and disaster contingency plans.

Uplift began again in 1992, although without a dramatic increase in earthquakes. Then, on September 18, 1994, two strongly felt quakes occurred nearly simultaneously beneath the caldera, producing a small

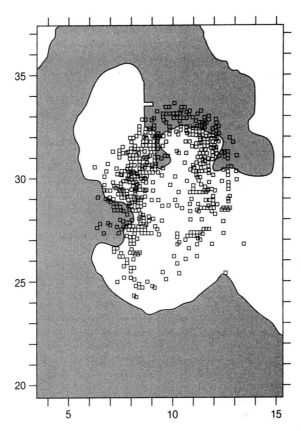

170   The locations of shallow earthquakes, most of them at depths of
3 kilometers or less, beneath Rabaul Caldera in Papua New Guinea from
September 1983 to November 1984. The area of maximum uplift was within the
oval formed by the earthquake locations. The numbers along the axes indicate
distances in kilometers. (From C. O. McKee, *Volcano News* 19–20 [1985]: 3.)

tsunami in the harbor. A large uplift occurred during the night of September 18–19, but was not detected until morning. By that time, Tavurvur
Volcano was already erupting. With only 27 hours of seismic warning,
the long-expected eruption had begun. Just over an hour after the start
of the Tavurvur eruption, Vulcan joined in from across the bay (Figure
171).

The combined eruptions dumped volcanic ash on Rabaul Town in
layers 10 centimeters to more than 1 meter thick. Pyroclastic flows from
Vulcan swept into the bay, generating tsunamis that reached as far as 200
meters onshore around parts of the harbor. Vulcan's eruption was over
in two weeks, but at Tavurvur, small, intermittent eruptions have continued for many months.

171 View of the Rabaul, Papua New Guinea, eruption from the Space Shuttle *Discovery* on September 19, 1994. The high eruption cloud from Vulcan Volcano reached an altitude of 18 kilometers. More than 50,000 people were evacuated from the area. (Photograph courtesy of NASA.)

Based on the earthquakes, a stage 2 alert (hazard situation possible) had been declared at 7:20 P.M. on September 18, eleven hours before the eruption began. A stage 3 alert (hazard situation probable) was declared only minutes before the eruption began. Stage 4 (evacuation required) was never announced, largely because most residents of Rabaul had begun to evacuate voluntarily during the night and early morning before the eruption started. Prompted by numerous sharp earthquakes, the majority of residents had fled to safer areas before 5:30 A.M. By 9:00 A.M., three hours after the eruption had begun, Rabaul was empty.

Destruction from ashfall was extensive in the city, with its south and east sides listed as 100 percent damaged. Where the ashfall was more than half a meter thick, almost all buildings collapsed, but where it was less than 20 centimeters thick, most roofs survived. Heavy rainfall on the loose ash caused lahars, and tsunamis severely damaged the wharf area. Pyroclastic flows devastated the area around Vulcan, but did not reach Rabaul. Luckily, the 750 people who lived near Vulcan had evacuated early.

Property damage from the 1994 Rabaul eruption totaled more than $200 million, but only eight people died, three of them in car accidents. With the short warning time and no mandatory evacuation, it was remarkable that the casualties were so few. Two factors seem to have been responsible: first, many older and respected people remembered the deadly 1937 eruption, and second, the 1984 earthquake crisis had not been forgotten, even though it had stopped without an eruption. People remembered the disaster plans and practice evacuations and fled without being ordered to do so.

After the immediate crisis of a volcanic eruption has passed, serious problems usually remain. Large numbers of people who have had to evacuate sometimes remain displaced for years, either because their homes have been destroyed or because continued volcanic activity threatens the evacuated area. Often large areas of croplands are destroyed, as happened in the Tambora eruption in Indonesia in 1815, after which starvation killed an estimated 80,000 people.

Even volcanologists fall victim to the uncertainties and risks of volcanic eruptions; 23 have been killed on volcanoes between 1979 and 2000. A small explosive eruption of Galeras Volcano in Colombia in 1993 killed six volcanologists who were in the summit crater taking gas samples. It is ironic that those killed were trying to better understand volcanic processes. Was this a case of "familiarity breeds contempt"? Perhaps, but it also led volcanologists to adopt many of the safety procedures that they advise for others. In hindsight, there were a few unusual local seismic events that preceded the explosion. These vibrations made a record on a recording drum called a *tornillo* (screw, in Spanish), a long-period signal lasting a minute or two with a sharp onset and a slowly decreasing amplitude that produces a record shaped like the side view of a screw. Such long-period seismic events are thought to be related to pressure changes in fluid flow at shallow depths.

One of the more notorious volcanic eruptions of the past decade has occurred at the Soufrière Hills Volcano in Montserrat, a Lesser Antilles island in the Caribbean. Steam and gas explosions began at this dormant volcano in July 1995 following increased seismic activity that began in 1992. These explosions occurred without ejecting any new solid magmatic material. In September 1995 a small lava dome of viscous andesite was extruded in the ancient crater. The rate of lava extrusion into the dome increased in January 1996, and the first pyroclastic flow resulting from dome collapse occurred in March 1996. A doubling of the extrusion rate to 4 cubic meters per second in July 1996 was followed by more frequent dome collapses and pyroclastic flows.

The extrusion rate continued to increase, and explosive magmatic eruptions began in September 1996. On June 25, 1997, a large dome collapse, explosive eruption, and pyroclastic flow killed 19 people. The eruption column reached a height of 12 kilometers. More pyroclastic flows destroyed Plymouth, the evacuated capital of Montserrat, in August 1997 (Figure 172). About 8000 residents left the island and about 4000 remained in its northern area, farthest from the volcano. Episodic dome growth, explosive eruptions, and pyroclastic flows have continued at Montserrat to the time of this writing (March 2005), with some periods

172   The Clock Tower in Plymouth, Montserrat, West Indies, has been buried by multiple pyroclastic flows. Plymouth had been evacuated prior to the first pyroclastic flow that destroyed the town in August 1997. (Photo taken by Montserrat Volcano Observatory in August 2001.)

of reduced activity. The latest major dome collapse occurred in July 2003. Activity since then has been at a relatively low level compared with earlier years.

Recent support for reducing volcanic risk has come from an unexpected source: the international airline industry. Near-fatal aircraft accidents resulted when jetliners temporarily lost power after flying into ash clouds from Galunggung Volcano in Indonesia in 1982 and Redoubt Volcano in Alaska. Redoubt had been quiet for 21 years, but on December 14 and 15, 1989, four major explosive eruptions shook the mountain out of its slumber. The largest jetting explosion began on the morning of December 15 and lofted an ash cloud to 40,000 feet. High-speed winds aloft carried the cloud to the north.

A Boeing 747 was approaching Anchorage on a polar flight from Amsterdam; it was 150 miles downwind from Redoubt when, at 25,000 feet in altitude, it encountered the deadly ash cloud, which was obscured in the general overcast. The pilot tried to climb, but it was too late; volcanic ash particles clogged and shut down all four engines. In eerie silence and darkness, the jumbo jet, with its 231 terrified passengers and crew, glided downward for eight minutes, losing 18,000 feet of altitude.

While the pilots struggled to restart the engines, the plane was coming perilously close to the mountains around Anchorage. Finally, with only 6000 feet and three minutes left, the engines shuddered back to life, and the plane made a miraculous safe landing.

Mechanics examining the engines later discovered the problem that had nearly downed the 747 and found out why the emergency measures had saved it. Ash particles that were swept into the hot, operating engines had melted into glassy coatings on the engines' turbines. Air intake was reduced by this clogging, and the engines shut down. As the plane glided beneath the ash cloud into clearer air, the glass coatings chilled and cracked, and were partially broken off by the repeated tries to restart the engines, which allowed air intake to be reestablished and saved the flight. Replacing the damaged engines and the "ash-blasted" windshields and leading edges of the wings, as well as a complete overhaul of the aircraft, cost 80 million dollars—a costly accident, but luckily, not a deadly one.

Since those near-fatal encounters, scientists and the international airline industry have joined forces to identify hazardous volcanic clouds as they form and to detour aircraft around them. A recent experiment by William Rose and David Schneider of Michigan Technological University indicates that volcanic ash clouds can be monitored from space. Using weather satellite observations of an ash eruption from Popocatépetl Volcano in Mexico in March 1996, they showed that by using images of different frequency bands from the satellite, they could discriminate between the ash cloud and weather clouds.

What will really reduce volcanic risk over the next century is not so much new concepts and techniques—although these will surely come about—but the application of currently known methods. Less than 10 percent of the world's 1300 potentially active volcanoes have been mapped to appraise their hazards, and even fewer are being monitored continuously. Educating the people at risk is also necessary; some people who live in Seattle will tell you that Mount Rainier is an extinct volcano (Plate 17). As this fourth edition of *Volcanoes* is being written, we have the ways to reduce volcanic risk, but not the means. Money to prevent catastrophes is always in short supply compared with money to repair their devastation.

But reducing volcanic risk is only part of understanding volcanoes. Volcanology is a young science. It has come a long way, but it has an even longer way to go on the path to a true understanding of how volcanoes work. This book is dedicated to those who will seek that path.

Links

Can Another Great Eruption Occur in Alaska?
http://geopubs.wr.usgs.gov/fact-sheet/fs075-98/

Colombian Volcanoes
http://vulcan.wr.usgs.gov/Volcanoes/Colombia/description_colombia_volcanoes.html

Domes of Destruction, NASA
http://earthobservatory.nasa.gov/Study/Domes/

Montserrat Volcano Observatory
http://www.mvo.ms

Redoubt Volcano, Alaska
http://www.volcano.si.edu/gvp/world/volcano.cfm?vnum=1103-03-

Smithsonian Global Volcanism Program, Soufrière Hills
http://www.volcano.si.edu/world/volcano.cfm?vnum=1600-05-

USGS Volcano Hazards Program
http://volcanoes.usgs.gov/

Volcanic Ash Aircraft Danger
http://pubs.usgs.gov/fs/fs030-97/

Volcano Hazards
http://pubs.usgs.gov/fs/fs002-97/

Location of Volcano Hazards
http://volcanoes.usgs.gov/Hazards/Where/WhereHaz.html

Sources

Blong, Russell, and C. McKee. *The Rabaul Eruption 1994.* Sydney: Natural Hazards Research Centre, Macquarie University, 1995.

Brantley, S. R., ed. *The Eruption of Redoubt Volcano, Alaska, December 14, 1989–August 31, 1990.* U.S. Geological Survey Circular 1061, 1990.

Bruce, V. *No Apparent Danger: The True Story of Volcanic Disaster at Galeras and Nevado del Ruiz.* New York: HarperCollins, 2001.

Casadevall, Tom. "Volcanic Ash and Aviation Safety." U.S. Geological Survey Bulletin 2047, 1994.

Fisher, R. V. *Out of the Crater.* Princeton, NJ: Princeton University Press, 1998.

Peterson, Donald, and R. I. Tilling. "Interactions Between Scientists, Civil Authorities and the Public at Hazardous Volcanoes." In C. R. J. Kilburn and G. Luongo, eds., *Active Lavas.* London: UCL Press, 1993.

Possekel, A. *Living with the Unexpected: Linking Disaster Recovery to Sustainable Development in Montserrat.* New York: Springer, 1999.

Rose, W. I., and D. J. Schneider. "Satellite Images Offer Aircraft Protection from Volcanic Ash Clouds." *Eos (Transactions, American Geophysical Union),* 77 (1996): 529–532.

Scarpa, R., and R. I. Tilling, eds. *Monitoring and Mitigation of Volcano Hazards.* Berlin: Springer-Verlag, 1996.

Tilling, Robert, and P. W. Lipman. "Lessons in Reducing Volcanic Risk." *Nature* 364 (1993): 277–280.

Williams, S., and F. Montaigne. *Surviving Galeras.* Boston: Houghton Mifflin, 2001.

CHAPTER 20

Volcanoes in the Solar System

173 Earth's full moon. The maria, the dark gray areas, are large impact basins filled with basaltic lavas. A large meteor crater near the bottom of the image with long rays of ejected material postdates the maria. (NASA.)

The full moon on a clear, warm night lures the mind to the mysteries that lie beyond the Earth. Can you see the "Man in the Moon"? His profile in dark gray may appear against the lighter gray background.

Early astronomers named the dark gray areas *maria,* the Latin name for seas, in the mistaken notion that they were somehow related to large bodies of water. The real men in the moon—the U.S. astronauts from the Apollo program who landed there—established that the maria were large basaltic lava flows that long ago filled huge meteor impact craters formed early in the moon's history.

Before the era of space exploration, moon-watchers with telescopes speculated on the origin of the thousands of craters that pock the lighter gray highlands surrounding the maria. Were they volcanic craters or impact craters? The space missions have clearly established that they are impact craters, most of them formed during the very early years of the moon's history.

With each new space exploration, it has become clearer that volcanism is not limited to the Earth, but has also been important on the Earth's moon, Mars, Venus, and the moon of Jupiter called Io. Space missions have established that past volcanic activity has made major contributions to the evolution of these bodies, and in the case of Io, that ongoing eruptions of lava and sulfur gases are constantly changing the face of this remarkable moon.

The Earth's moon has a radius 0.27 times that of the Earth. Present theory considers it to be the result of the collision of another body with the early Earth. After its formation by accretion of the collision fragments, the moon had a surface layer of magma that cooled into its present crust. During and shortly after this solidification, major meteor impacts created the broken and densely cratered surface still seen in the present highlands (the lighter gray areas in Figure 173). During or shortly after this period, some very large impacts with objects as much as 100 kilometers in diameter gouged out the maria basins. Large flows of basaltic lava then filled these basins to form the present maria.

Most of the Apollo moon landings were on the relatively smooth surfaces of the maria, and the astronauts returned with basalt samples that ranged from 4.3 to 3.1 billion years in age. The basaltic lava is fine-grained from rapid cooling. Its very low eruption viscosity has created some flows as much as 800 kilometers long, 20–40 kilometers wide, and 50 meters thick. Small vesicles occur in a few samples, but there are no signs of any minerals that could have contained water. Carbon monoxide has been suggested as the gas that formed the bubble holes. The age of the youngest lavas is not known, but the low crater densities seen on images of some maria flows suggest a lower age limit of about 1 billion years. The dark gray maria basalts cover about 20 percent of the moon's surface, but form less than 1 percent of the moon's volume.

The fact that the moon is no longer volcanically active agrees with the general principle that the smaller a planet or moon, the shorter its expected volcanic life span should be. Uranium content, heat conductivity, and other factors being roughly equal, large bodies generate heat at a rate proportional to the cube of their radius and lose surface heat at a rate proportional to the square of their radius; hence small, rocky planets should cool more rapidly than large ones.

The astronauts also examined the Hadley Rille (Figure 174 and 175), one of the largest of the many sinuous rilles that appear on the maria. These rilles are generally considered to have formed as lava channels or collapsed lava tubes. The Hadley Rille as seen on flyby images is 100 kilometers long, 1–3 kilometers wide, and as much as 1 kilometer deep.

Tiny beads of black, orange, and green basaltic glass occur in many of the loose rock samples collected by the astronauts. These are considered to be the moon's equivalent of the fine spray of lava fragments formed at lava fountains on the Earth. In the moon's lower gravity, these glass fragments are widely dispersed and do not build up the classic cinder and spatter cones seen in Hawaii. Vents related to the eruption of the maria basalts are considered to be the principal sources of these pyroclastic glasses.

Mars has a radius 0.53 times that of Earth. Several unmanned spacecraft missions to Mars, beginning in 1971, have revealed spectacular images of many volcanoes, including the giant shield volcano Olympus Mons, the largest volcano in our solar system (Figure 176). At 600 kilometers in diameter and 25 kilometers high, Olympus Mons dwarfs Mauna Loa, Earth's largest volcano; in fact, the entire Hawaiian Islands chain would fit within its diameter (Figure 177). The caldera on Olympus Mons is 85 by 65 kilometers across and 3 kilometers deep. The Hawaiian island of Maui would fit inside this vast caldera.

174 Hadley Rille, an ancient lava channel, as seen from the lunar lander on the *Apollo 15* mission, NASA. (NASA.)

The age of the Martian volcanoes is less certain than the age of those on the moon because no samples have yet been returned from Mars for analysis. However, the density of meteor impacts on Mars (high early in Mars history; low in more recent Mars history) does provide some clues. Olympus Mons and the other four great shield volcanoes of the Tharsus region appear to have mainly grown about 1 to 2 billion years ago, but their youngest flow ages are not known. Recent investigations by the European Space Agency suggest that the youngest caldera floor flows on Olympus Mons may be less than 2 million years old (Figure 178).

Lava flows from Arsia Mons Volcano were apparently of very low viscosity, traveling as far as 800 kilometers. Many investigators infer that these flows are basaltic in composition, but without samples to analyze, this is still just an educated guess. The history of space exploration includes as many surprises as good guesses. About 60 percent of Mars's sur-

175 *Apollo 15* astronaut Alfred Worden and the lunar rover on the edge of Hadley Rille. (NASA.)

face is plains that could be lava flows similar to the maria on the moon. However, they could also be alluvial or windblown deposits.

Twenty-three major volcanic structures ranging from 3 to 600 kilometers in diameter have been given formal names on Mars, and there are hundreds of smaller features that appear to be of volcanic origin. When volcanologists finally roam on Mars, there will be plenty to keep them busy!

Venus has a radius nearly (0.97 times) as large as Earth, and it was anticipated that a planet this size should have a volcanic history. The problem was how the surface of Venus could be seen from spacecraft. Russian probes that landed on the surface of Venus in the 1980s had established that its surface temperature is 470°C and that the pressure of its 97 percent carbon dioxide atmosphere is 90 times greater than the Earth's atmospheric pressure. The final hurdle to overcome was the complete cover of sulfuric acid clouds at an elevation of 45 to 60 kilometers above the ground surface, which prevents any optical photography.

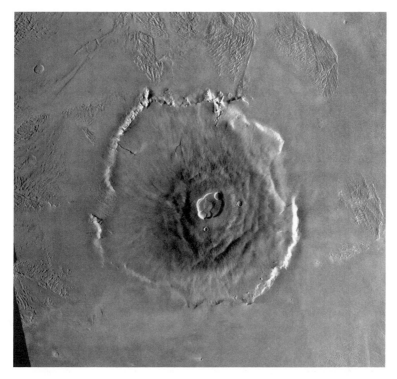

176 Olympus Mons on Mars, the largest volcano in our solar system, is 600 kilometers in diameter and 25 kilometers high. Its caldera is 85 by 65 kilometers across and as much as 3 kilometers deep. (NASA.)

The answer was radar, and the Magellan mission in the early 1990s made spectacular radar maps of almost the entire surface of Venus. Volcanoes and strange tectonic features were the predominant landforms revealed by these orbiters. Scientists have cataloged more than 1700 volcanic landforms, including 168 large volcanoes more than 100 kilometers in diameter and many of intermediate size (from 20 to 100 kilometers in diameter). Specific volcano types include effusive shields and low to steep-sided volcanoes similar to but much larger than volcanic domes on Earth (Figure 179), as well as calderas and complex volcanic features. In addition, lava plains and lava flow fields cover much of Venus's surface.

There is no evidence of extensive pyroclastic flow deposits in the radar images. It is probable that the high atmospheric pressure on Venus—equal to the pressure at a depth of 1 kilometer beneath the ocean on Earth—retards explosive eruptions.

The Russian landers identified most of the surface composition of Venus as basalt and material with several percent sulfur. Radiometric

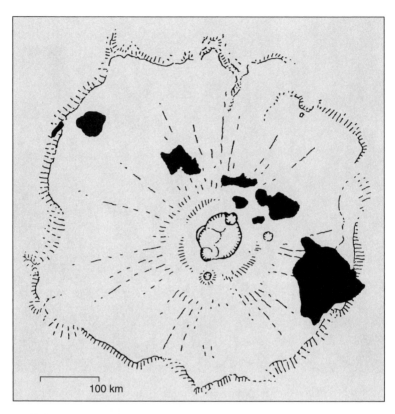

177 All the major islands of the State of Hawaii would fit inside the diameter of Olympus Mons Volcano on Mars. (Drawing by Rick Hazlett, used with permission.)

ages of the volcanoes and lava plains are not known, but the low impact crater densities on their surfaces imply that they are young relative to the overall age of Venus. Planetary scientists studying this problem think that a major "resurfacing event" occurred on Venus sometime between 1600 and 300 million years ago.

There is no stream erosion and little wind erosion on this sister planet to Earth. Water vapor is converted to sulfuric acid, and the high atmospheric density retards wind currents. The volcanic landforms are beautifully preserved, but their ages are uncertain. Some students of Venus believe that the planet undergoes episodic outbursts of global volcanism, followed by periods of relatively low volcanic activity.

On Earth, water, calcium from the chemical erosion of rocks, and carbon dioxide from volcanoes are consumed by living reefs that build great banks of limestone—calcium carbonate. This mechanism retards the buildup of carbon dioxide, a "greenhouse gas," in our atmosphere.

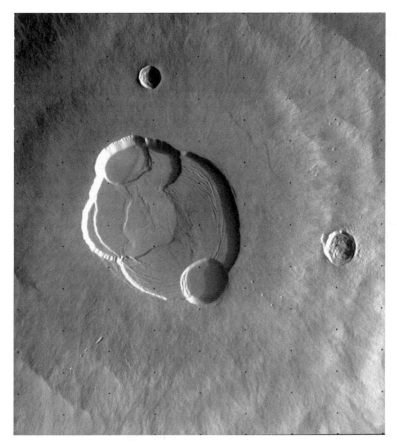

178 The huge caldera of Olympus Mons Volcano on Mars shows multiple episodes of collapse. The lack of impact craters on the lava flows ponded in the caldera floor has led to speculation that Olympus Mons is still a potentially active volcano. (NASA.)

The dense carbon dioxide atmosphere and high surface temperature of Venus suggest an alternative planetary evolution. Is this fact or speculation? Venus, the Goddess of Love, still hides many mysteries beneath her veil of clouds.

Io, the innermost moon of Saturn, has a radius 0.29 times that of Earth. Though small, it has turned out to be the most volcanically active body in our solar system. The story began in 1979 as scientists at the Jet Propulsion Laboratory were processing the first images of Jupiter and its moons from the spacecraft *Voyager 1*.

Linda Morabito was intensifying an image of Io in order to get an exact fix on two faint stars behind the moon. She not only obtained the

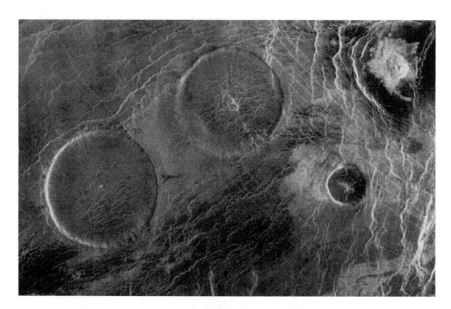

179 Radar images of Venus from the *Magellan* spacecraft show many features of volcanic origin. Among them are these dome-shaped hills about 25 kilometers wide and 750 meters high. Because of their flat tops, these apparent lava domes have been informally named pancake domes. (NASA.)

stars' locations, but also noticed a strange image on the surface of Io that reached upward 280 kilometers. She considered that this strange image might be a plume of gas or particles from an active volcano erupting into Io's low gravity. She showed the image to her colleagues, and they agreed that it represented a historic discovery.

A few months before *Voyager 1* reached Jupiter, S. J. Peale, P. Cassen, and R. T. Reynolds, scientists at the University of California and NASA, predicted that Io might have a molten interior caused by tidal friction. They noted that the orbit of Io was distorted by the great tidal strains from Jupiter and its neighboring moons, and speculated that "widespread and recurrent surface volcanism would occur." The prediction of active volcanism on Io from theoretical calculations, and its subsequent observation on the flyby images, is an example of scientific research at its best.

The flyby of *Voyager 2*, only 18 weeks after *Voyager 1*, confirmed Io's intense volcanic activity and showed major changes in its volcanic deposits (Plate 21). In addition to the dramatic color images of apparent sulfur deposits—black, red, orange, and yellow—more plumes were discovered, and the infrared interferometer spectrometer detected many areas of high thermal emission. Io's heat flow is 10^{14} watts, compared with Earth's 4.4×10^{13} watts, some 200 times the amount of heat flow expected from the radioactive elements in Io's interior. Flexing of Io's crust and internal layers by the large and changing gravitational pull of Jupiter and its adjacent moons maintains this great heat flow and ongoing volcanic activity.

Although much of the surface of Io is dominated by smooth volcanic plains, there are some calderas about 20 to 200 kilometers wide and as much as 2 kilometers deep. There are also some blocklike mountains as high as 17 kilometers. Sulfur deposits are too weak to support such rugged topography, so it was suspected that silicate lavas must also be present on Io. Recent infrared measurements have confirmed this idea, showing lava flow temperatures as high as 1500°C.

Io's volcanoes are very young, ranging from ongoing eruptions to cooled but young structures without any signs of impact craters. More than 350 volcanic centers have been identified; not all of them are active, but they are widely distributed. Named volcanic centers such as Pele and Prometheus have emitted observed eruption plumes of gases and particles; Pele's plume has risen as high as 460 kilometers. Both the low gravity of Io and its near-zero atmospheric pressure contribute to the formation of these high plumes (Figure 180).

Mercury, the planet closest to the sun, has a radius 0.38 times that of Earth. It has been explored by one photographic mission in 1974–1975, which obtained only partial coverage of its surface. In addition to its heav-

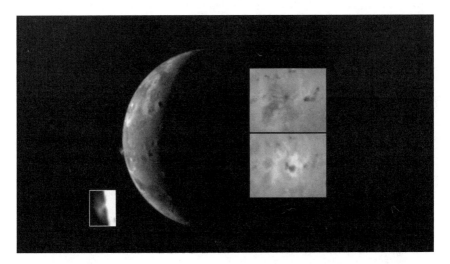

180 Two volcanic plumes on Io. The 86-kilometer-high plume on the left is erupting from a caldera called Pillan Patera. The other, seen near the sun's shadow, is erupting from a volcanic center called Prometheus; it is possible that this plume has been active for more than 18 years. (NASA.)

ily cratered terrain, Mercury has regions of relatively smooth plains, some of which may be the result of ancient volcanic activity. Reanalysis of the *Mariner 10* data in 1997 indicates that both lava flows and pyroclastic deposits may account for some of the features seen on the images.

During the 1963–1972 Apollo moon missions and many subsequent space missions, a great deal has been learned about the volcanism of the planets and larger moons in our solar system. Volcanism has turned out to be a dominant geologic process in their formation and evolution. The similarities and differences among volcanic features and processes on Earth and on those other planets and moons are of great importance in understanding their fundamental nature. For example, the great size of Olympus Mons is apparently related to the lack of plate tectonic features on Mars and the long-term eruption of a "hot spot" without plate motion spreading it out into a long chain of volcanoes. The volume of the entire Hawaiian Ridge–Emperor Seamount chain is comparable to the volume of Olympus Mons.

Fifty years of space exploration has created a new science of comparative planetary volcanology. It is only a beginning. What lies beyond all these new mountains?

Links

Volcanism on the Moon
http://volcano.und.nodak.edu/vwdocs/planet_volcano/lunar/Overview.html

Volcanoes in Space
http://www.earthsky.com/shows/showsmore.php?t=20031007

Apollo Program
http://nssdc.gsfc.nasa.gov/planetary/lunar/apollo.html

Moon Thumbnails
http://nssdc.gsfc.nasa.gov/imgcat/thumbnail_pages/moon_thumbnails.html

Olympus Mons
http://en.wikipedia.org/wiki/Olympus_Mons

Geology of Mars: Volcanoes
http://aerospacescholars.jsc.nasa.gov/CAS/lessons/L9/10.HTM

Mars Recent Eruptions
http://www.msnbc.msn.com/id/6746297/

Volcanoes on Venus
http://volcano.und.nodak.edu/vwdocs/planet_volcano/venus/intro.html

Magellan Venus Factsheet
http://www2.jpl.nasa.gov/magellan/

Io Volcanoes
http://science.nasa.gov/headlines/y2000/ast19may_2.htm

Eruption on Io
  http://antwrp.gsfc.nasa.gov/apod/ap000606.html

Sources

Beatty, J. K., C. C. Petersen, and A. Chaikin, eds. *The New Solar System*. 4th ed. Cambridge: Cambridge University Press, 1999.

Crumpler, L. S., and Jayne C. Aubele. "Volcanism on Venus." In H. Sigurdsson, ed., *Encyclopedia of Volcanoes*. San Diego, CA: Academic Press, 2000.

Lopes-Gautier, Rosaly. "Volcanism on Io." In H. Sigurdsson, ed., *Encyclopedia of Volcanoes*. San Diego, CA: Academic Press, 2000.

Magellan Science Team. "Magellan at Venus." *Journal of Geophysical Research* 97 (1992):.

McEwen, A. S. et al. "Very high temperature volcanism on Jupiter's moon Io." *Science* 281 (1998): 87–90.

Plummer, Charles C., David McGeary, and Diane H. Carlson. *Physical Geology*. 10th ed. New York: McGraw-Hill, 2005.

Robinson, Mark S., and Paul G. Lucey. "Recalibrated Mariner 10 Color Mosaics: Implications for Mercurian Volcanism." *Science* 275 (1997): 197–200.

Roth, L. E., and S. D. Wall, eds. *The Face of Venus*. NASA Special Publication SP-520. Washington, DC: National Aeronautics and Space Administration, 1995.

Spudis, Paul D. "Volcanism on the Moon." In H. Sigurdsson, ed., *Encyclopedia of Volcanoes*. San Diego, CA: Academic Press, 2000.

Wilhelms, D. E. *The Geologic History of the Moon*. U.S. Geological Survey Professional Paper 1348, 1987.

Zimbelman, James R. "Volcanism on Mars." In H. Sigurdsson, ed., *Encyclopedia of Volcanoes*. San Diego, CA: Academic Press, 2000.

Appendix A

The World's 101 Most Notorious Volcanoes

Activity, size, shape, beauty, danger, location, and the authors' preferences were used to select the following list. More complete catalogs of active volcanoes, from which some of the data in this appendix are derived, can be found in the Web sites listed in Appendix B. The eruption data are through 2004.

ANTARCTICA

Mount Erebus A snow- and ice-covered stratovolcano with an active lava lake in its summit crater. It was in eruption when first sighted in 1841 and has had several reported explosions since. The present lava lake, about 100 meters in diameter, apparently formed in the 1960s. Erebus is the world's southernmost active volcano. The last eruption was in 2004.

ATLANTIC OCEAN

Beerenberg A stratovolcano with a caldera, on Jan Mayen Island north of Iceland. It is the northernmost active volcano and has erupted five times since 1633, the last time in 1985.

Fogo A stratovolcano with an 8-kilometer-wide caldera, in the Cape Verde Islands. It was continuously active from the time of settlement, in about 1500, to 1760. Ten eruptions have occurred since then, the last in 1995. That eruption covered 6 square kilometers with about 80 million cubic meters of lava, destroying valuable farmland.

CARIBBEAN SEA

Mont Pelée A stratovolcano with summit domes, on Martinique. The explosive eruption on May 8, 1902, generated a *nuée ardente* that swept down the mountainside and, within minutes, incinerated the town of St. Pierre and its 28,000 inhabitants (see Figures 84 and 85). Three other explosive eruptions have been recorded, the last in 1932.

La Soufrière of Guadeloupe A stratovolcano with a summit dome, on the southern part of Guadeloupe Island. It has erupted explosively about ten times since 1400. The last eruption in 1976 prompted the evacuation of

70,000 people for several months, but the eruption turned out to be only minor explosions.

La Soufrière of St. Vincent A stratovolcano with a crater lake. Its seven eruptions since 1718 include two major explosive events. Evacuation in 1979 prevented a potential repeat of the tragedy of 1902 when 1600 deaths occurred, largely from *nuées ardentes*.

Soufrière Hills of Montserrat A stratovolcano with lava domes on the island of Montserrat. The first recorded eruption of this volcano began in July 1995. The growing lava dome has collapsed and exploded several times, producing pyroclastic flows. More than 6000 people were evacuated from the island. This is a dangerous eruption, and the remaining residents of the island remain on alert. The last eruption was in 2004.

CHILE

Cerro Hudson A stratovolcano in southern Chile. Known eruptions occurred in 1891 and 1971, and there was a major explosive eruption in 1991. The region near the volcano is sparsely populated, so no people were killed by the last eruption, but thousands of sheep and cattle died from fluorine poisoning from the ash-covered grasslands.

Llullaillaco A stratovolcano in northern Chile. It is the world's highest active volcano, with a summit elevation of 6739 meters. Of the three recorded eruptions during the 1800s, two were explosive, and the other was a lava flow from a side vent. Nevado Ojos del Salado (6886 meters), also in Chile, is higher, but has only steam vents and no recorded eruptions.

Villarrica A stratovolcano in central Chile. About 21 explosive eruptions and three lava flows have occurred since 1558. The eruption in 1971 melted large volumes of snow and ice, and the resulting mudflows killed 15 people. The last eruption was in 2004.

COLOMBIA

Galeras A complex volcano in southern Colombia. It has erupted more than 20 times since 1535. The explosive eruption in 1993 was small but killed nine people, including six volcanologists who were in the crater, making it clear that forecasting volcanic eruptions still has a long way to go. The last eruption was in 2004.

Nevado del Ruiz A high (5321 meters) stratovolcano in central Colombia. Mudflows from an earthquake or small eruption in 1845 killed about 1000 people. Another small to moderate eruption in 1985 melted part of the summit ice cap; the resulting mudflows caused about 23,000 fatalities (see Chapters 10 and 19). The last eruption was in 1991.

COSTA RICA

Arenal A stratovolcano in northwestern Costa Rica. Dormant until 1968, it burst into eruption with a strong explosion, throwing huge blocks as far as

5 kilometers; the explosion was followed by *nuées ardentes*. This initial eruption killed 78 people. Activity has continued since, with the extrusion of thick, slow-moving lava flows. The last eruption was in 2004.

Irazú A stratovolcano with a double crater, in central Costa Rica. It has erupted explosively about 15 times since 1723. A major eruption lasted from 1963 until 1965; its numerous small to moderate ashfalls were destructive to coffee plantations and a nuisance to the capital city of San Juan. The last eruption was in 1994.

Poás A stratovolcano with twin crater lakes, in central Costa Rica. It has erupted 24 times since 1834, generally with explosions of mud and water from the northern lake. The 1910 eruption shot a fountain of water more than 4 kilometers high. The last eruption was in 1996.

DEMOCRATIC REPUBLIC OF THE CONGO

Nyamuragira A shield volcano with a summit caldera, in eastern Congo. It has erupted more than 20 times since 1894, including lava lake activity from 1921 to 1938. The last eruption was in 2002.

Nyiragongo A stratovolcano with a summit caldera, in eastern Congo. It has erupted about 15 times since 1884, including lava lake activity from 1935 to 1977. A major fissure eruption on the south flank in 1977 drained the lava lake and rapidly covered an area of several square kilometers with very fluid lavas. About 300 people were killed by these flows. The last eruption was in 2004.

ECUADOR (ANDES)

Cotopaxi A stratovolcano nearly 6000 meters high (see Figure 98). It has had more than 50 eruptions since 1532, including major eruptions and lava flows. The 1877 eruption melted large volumes of snow and ice from the summit, causing mudflows that traveled 100 kilometers down adjacent riverbeds. The last eruption was in 1940.

Guagua Pichincha A stratovolcano with a caldera and a central cone. Although it was active from the 1500s to the 1800s, there have been only small eruptions since, the last one in 2004. Forty centimeters of ash fell on the capital city of Quito in 1660; modern Quito climbs up the sides of this potentially dangerous volcano.

Reventador A stratovolcano east of the main ranges of the Andes. Although this mysterious volcano was not explored until 1931, it was apparently the source of some ashfalls in Quito as far back as 1541. An important oil pipeline connecting the Amazon fields to the west coast of Ecuador now crosses its north slope. The last eruption was in 2003.

Sangay A 5230-meter-high snow- and ice-covered stratovolcano. Small explosive eruptions with occasional lava flows kept this volcano in nearly continuous activity from 1728 to 1916 and from 1934 through 2004.

Tungurahua A stratovolcano that rises in a sharp cone 3 kilometers above its base. It is one of Ecuador's most active volcanoes. Eruptions of pyroclastic flows and lava flows have reached populated areas near the volcano's base. The recent long-term eruption that began in 1995 brought about a temporary evacuation of the city of Banos. The last eruption was in 2004.

ECUADOR (GALÁPAGOS ISLANDS)

Fernandina A shield volcano with a summit caldera. The most active of the Galápagos volcanoes, this uninhabited island was called Narborough in Darwin's chronicles. A major explosive eruption and 350-meter collapse of the caldera occurred in 1968. The last eruption was in 1995.

EL SALVADOR

Izalco A young stratovolcano born in 1770 on the south flank of Santa Ana Volcano in western El Salvador. It had nearly continuous small explosive eruptions until 1957 and was known as the Lighthouse of the Pacific. When a hotel was built nearby to view the frequent eruptions, the activity stopped. The last eruption was in 1966.

ETHIOPIA

Erta Ale A shield volcano with an active lava lake, in the rift valley of northern Ethiopia. Lava eruptions from fissures on the flank were observed in 1959–1960. The active lava lake discovered in 1967 has since been in nearly constant eruption.

GREECE

Santorini A stratovolcano with a submerged caldera, in the Aegean Sea. Its giant explosive eruption and caldera collapse in about 1600 B.C. buried Akroteri, an important Minoan city currently under excavation. The huge eruption and sudden sinking of the island's center beneath the sea may have been the source of the legend of Atlantis. The last eruption in 1950 formed a lava dome and thick lava flows on the islands within the caldera.

GUATEMALA

Atitlán A stratovolcano on the south rim of a caldera lake 20 kilometers in diameter, in southwestern Guatemala. It has erupted with small to moderate explosions 11 times since 1469. The caldera lake, formed following immense prehistoric eruptions, is one of the most beautiful in the world. The last eruption was in 1853.

Fuego A stratovolcano—whose name means "fire"—in southwestern Guatemala. It has erupted more than 50 times since 1524, producing mostly explosions of ash, but sometimes *nuées ardentes* and lava flows. The last eruption was in 2004.

Pacaya A volcanic complex of two small stratovolcano cones and older lava domes, in southern Guatemala. It has erupted more than 40 times since 1565, generally only with explosions but with some lava flows in recent years. The last eruption was in 2002.

Santa Maria A stratovolcano with a growing lava dome on its southwest slope, in western Guatemala. Its first known eruption was in 1902; it produced a giant explosion of 5.5 cubic kilometers of pumice fragments and ash. A lava dome named Santiaguito began growing in the explosion crater in 1922 and has since been erupting intermittently. The last eruption was in 2004.

ICELAND

Askja A complex volcano with a caldera 10 kilometers in diameter, in central Iceland. Although most of its eight eruptions since the fourteenth century have been lava flows, the great explosion of 1875 showered 2 cubic kilometers of ash over much of eastern Iceland. The resulting famine led many Icelanders to emigrate to the United States and Canada. The last eruption was in 1961.

Grímsvötn A caldera in the Vatnajökull ice cap of south central Iceland. It has erupted beneath the ice about 25 times since 1332, causing gigantic floods called *glacial bursts*. The sudden floods often exceed the flow volume of the Mississippi River. The last eruption was in 1998.

Heimaey A cinder cone with a thick, blocky, lava flow from its north side, in the Vestmann Islands off the south coast of Iceland. A 2-kilometer-long fissure opened near the fishing port of Heimaey in 1973. Cinders and ash soon covered much of the evacuated town of 5000 inhabitants, and the thick flow nearly closed off the harbor entrance. The courageous Icelanders returned to a better harbor, rebuilt their town, and even heated their hospital with steam from the cooling flows.

Hekla A stratovolcano elongated by a northeast-trending rift system, in south central Iceland. It has erupted 20 times since the settling of Iceland in A.D. 900, generally with ash explosions followed by lava flows. In medieval Europe, Hekla was considered the gate to hell. The last eruption was in 2000.

Krafla A complex volcano with a large central caldera, in northern Iceland. Dormant after a series of eruptions between 1724 and 1728, Krafla awakened in 1975 with small eruptions and episodes of extensive ground cracking. An eruption in 1977 also sprayed a small amount of lava out of a producing geothermal steam well, the only known case of an eruption from a man-made vent (see Figure 30). The last eruption was in 1984.

Laki A fissure zone more than 25 kilometers long, in south central Iceland (see Figures 89 and 149). Its single giant eruption in 1783 produced 15 cubic kilometers of lava, a historic record, filling two river valleys and covering more than 500 square kilometers. Stunted grass and fluorine poisoning resulting from the accompanying volcanic gases starved and killed most of Iceland's livestock. The ensuing famine caused 10,000 deaths.

Surtsey A cinder cone and lava flow island on the south coast of Iceland. Born from the sea in 1963 and erupting until 1967, it has provided scientists a view of how land forms and plants and animals establish themselves in this new territory (see Chapter 2).

INDIAN OCEAN

Piton de la Fournaise A shield volcano with a caldera, on the eastern part of Réunion Island. Sometimes called a sister to Hawaiian volcanoes because of their similarity of climate and volcanic nature, it has erupted lava flows more than 100 times since 1640. The last eruption was in 2004.

INDONESIA

Agung A stratovolcano in Bali, considered in legend to be the navel of the world. Although it has erupted explosively only four times since 1808, the last eruption in 1963–1964 was of major proportions. High ash explosions affected world climate, and pyroclastic flows killed 2000 people.

Colo A low (508 meters) stratovolcano with a 2-kilometer-wide caldera and small crater lake, on Una Una Island near Sulawesi (Celebes). A large explosive eruption occurred in 1983 just after the 7000 inhabitants of the island had been evacuated (see Chapter 19).

Dukono This complex volcano in northern Halmahera is one of Indonesia's most active volcanoes. Nearly continuous explosive eruptions, sometimes accompanied by lava flows, have occurred from 1933 through 2004, and are continuing.

Galunggung A stratovolcano with a lava dome, in western Java. It has erupted only five times in recorded history, but the first eruption in 1822 produced a 22-kilometer-long debris avalanche and mudflow that killed 4000 people. The 1982 eruption killed 27 people. The last eruption was in 1984.

Gamalama A stratovolcano island with multiple craters and crater lakes, west of Halmahera. It has erupted explosively more than 60 times since 1538, sometimes producing lava flows. The last eruption was in 2003.

Kelut A stratovolcano with a crater lake, in eastern Java. It has erupted 30 times since about A.D. 1000. The explosive eruptions eject the hot crater lake and cause widespread destruction. In the 1919 eruption, more than 100 villages were destroyed or damaged by mudflows that killed 5100 people. The last eruption was in 1990.

Krakatau Stratovolcano islands around a submerged caldera, in the Sunda Strait between Sumatra and Java. The 1883 eruption was one of the largest natural explosions in recorded time. Sounds were heard for 4000 kilometers; the emitted ash and pumice blocks totaled 18 cubic kilometers; the 6-kilometer caldera collapsed; and the resulting tsunamis killed 36,000 people on the low shores of Java and Sumatra. The last eruption was in 2001.

Merapi A stratovolcano—whose name means "mountain of fire"—with a summit lava dome, in central Java. It has erupted more than 60 times since A.D. 1006, generally with explosions and *nuées ardentes*. The 1006 eruption caused so much death and destruction that the Hindu rajah moved to Bali, and Java became Muslim. The last eruption was in 2002.

Semeru A stratovolcano with a summit lava dome, in eastern Java. Its nearly 70 eruptions—generally explosions sometimes accompanied by pyroclastic flows and lava flows—make it one of Java's most active volcanoes as well as its highest (3620 meters). The last eruption was in 2004.

Tambora A stratovolcano with a summit caldera, on Sumbawa Island. Its giant eruption in 1815 exceeded the size and power of Krakatau's. The explosion, followed by caldera collapse, is estimated to have produced about 40 cubic kilometers of ash and blocks. Ten thousand people were killed by the eruption, and 80,000 starved in the resulting crop loss and famine. World climate may have been affected (see Chapter 17). The last eruption was in 1967.

ITALY

Etna A transitional shield-to-stratovolcano in northeastern Sicily. It has erupted lava flows more than 150 times since activity was first recorded in 1500 B.C. Small to moderate explosive eruptions occur at the summit; one in 1979 took nine lives. Major lava flows occurred during 2001 and 2002. The last eruption was in 2004.

Stromboli A stratovolcano island west of Italy. Known as the Lighthouse of the Mediterranean, it has been in almost continuous eruption for more than 2000 years. Small explosions of incandescent lava hurled up from the crater every 15 to 30 minutes are visible to ships passing by. Larger eruptions, some with lava flows, take place every few years.

Vesuvius A complex stratovolcano east of Naples. Most famous for its A.D. 79 eruption that buried Pompeii, it has since erupted more than 50 times. Explosive eruptions are generally followed by lava flows. The last eruption was in 1944 (see Figure 148).

Vulcano A stratovolcano west of Italy. Legendary forge of Vulcan, this small island has provided the family name for all volcanoes. It has erupted explosively about ten times since 200 B.C., the last time in 1890.

JAPAN

Asama A complex stratovolcano in central Japan. It has erupted more than 100 times since A.D. 685, generally with a series of small explosions. In

1783 large pyroclastic flows and mudflows buried villages, killing 1300 people. Scientists at the Asama Volcano Observatory, established in 1909, have found earthquake counts and locations to be useful in forecasting eruptions. The last eruption was in 2004.

Aso A group of cinder cones and small stratovolcanoes within a caldera 20 kilometers in diameter. One vent has erupted more than 100 times since A.D. 796, generally as single, isolated explosions. Tourists visiting the rim of the active vent are sometimes killed by ejected blocks and bombs. The last eruption was in 2003.

Bayonnaise Rocks Lava domes that sometimes build ephemeral islands inside a submarine caldera southeast of Japan. This submarine volcanic center has erupted more than ten times since it was first witnessed in 1896, sometimes building an island up to 95 meters high. A Japanese oceanographic ship investigating the area was destroyed by an explosion in 1952, with the loss of all 31 people on board. The last eruption was in 1988.

Fuji A classic stratovolcano in central Japan, the archetype of volcanic form (see Figure 57). It has erupted ash and lava about 15 times since A.D. 781. The last eruption in 1707 was from a vent high on the southeast side that ejected 0.8 cubic kilometer of ash, blocks, and bombs. The finer ash reached Tokyo.

Miyakejima A stratovolcano on an 8-kilometer-wide island in the Izu chain, 200 kilometers south–southwest of Tokyo. A major eruption began in July 2000 with lava flows and ashfalls, followed by large emissions of SO_2 gas. By September the 4000 inhabitants of the island had been evacuated. Although the gas emissions have diminished, they are still noxious, and the eruption and evacuation have continued into 2004.

Oshima A stratovolcano island with a caldera and a central cone, off the east coast of central Japan. It has erupted about 50 times since A.D. 684, sometimes explosively and sometimes with extensive lava flows (see Figure 81). The last eruption was in 1990.

Sakurajima A stratovolcano forming a peninsula into Kagoshima Bay, in southern Japan. One of the world's most active volcanoes, it has had thousands of small explosive eruptions since the first recorded one in A.D. 708. Many deaths occurred in 1476 and 1779, but sufficient warning from earthquakes kept the death toll to only a few people in the giant eruption of 1914, which produced 0.6 cubic kilometer of ash and 1.6 cubic kilometers of thick lava flows. The last eruption was in 2004.

Unzen A complex volcanic peninsula formed by several lava domes, on the west coast of southern Japan. Although only six eruptions have been recorded since A.D. 860, the 1792 activity involved either an explosion or an earthquake that triggered a 0.5-cubic-kilometer avalanche. The slide and resulting tsunami caused 14,000 deaths. The 1991 eruption killed 43 people, including three volcanologists (see Figure 166). The last eruption was in 1996.

Usu A stratovolcano with summit and flank lava domes, in northern Japan. Although it has erupted only seven times since 1663, most of these eruptions have been explosive and destructive to life or land. In 1910, 1943–1944, and 1977–1982, large lava domes or ground surface uplifts—kilometers in diameter and tens to hundreds of meters high—were slowly forced upward. The last eruption was in 2001.

MEXICO

El Chichón A small stratovolcano in Chiapas in southern Mexico, dormant until 1982. Its sudden explosive eruption generated pyroclastic flows that killed about 2500 people and injected a huge cloud of dust and sulfuric acid aerosol into the stratosphere (see Chapter 16).

Colima A stratovolcano with a summit lava dome, in west central Mexico. Its more than 40 eruptions since 1576 have been mostly explosive, with some lava flows. The last eruption was in 2004.

Parícutin A cinder cone in central Mexico. Born in a cornfield in 1943, it built a 410-meter-high cone with extensive lava fields during its brief life span. Most of the 1.3 cubic kilometers of ash and cinders and much of the 0.7 cubic kilometer of lava were produced in the first few years. The eruption ended in 1952.

Popocatépetl A snow-capped stratovolcano dominating the skyline south of Mexico City. It erupted with small explosions 11 times between 1512 and 1697. Since a major explosion in 1720, only five small eruptive episodes have occurred, the last in 1996–2004.

NEW ZEALAND

Ngauruhoe A stratovolcano with a nearly perfect cone, in central North Island (see Figure 111). It has had more than 50 explosive eruptions since 1839, some with small pyroclastic flows. The latest eruption was in 1977.

Ruapehu A stratovolcano with a hot crater lake, in central North Island (see Figure 111). There have been more than 30 small explosions of steam and ash from the crater lake since 1861. Mudflows from the erupting lake occasionally flood adjacent valleys. One such flow on Christmas Eve in 1953 swept away a railroad bridge, wrecking the Wellington-Auckland Express and causing 151 deaths. The latest eruption was in 1997.

Tarawera A volcanic complex of rhyolitic domes, in central North Island. A great eruption along a 17-kilometer-long fissure in 1886 ejected 1.3 cubic kilometers of basaltic ash and hot mud, burying three villages and killing more than 150 people. From 1900 to 1904, Waimangu ("black water") Geyser erupted in one of the craters formed along the 1886 fissure. Occasional geyser bursts of 400 to 500 meters during that time were the highest ever recorded. The last eruption was in 1973.

White Island A stratovolcano in the Bay of Plenty north of New Zealand, with a large horseshoe-shaped crater. About 35 small to moderate explosive eruptions have occurred since 1826. Eleven men were killed and a sulfur works was destroyed in the 1914 eruption and landslide. The last eruption was in 2001.

NICARAGUA

Cerro Negro A cinder cone—whose name means "black hill"—in western Nicaragua (see Figure 52). Born in 1850, it has erupted about 20 times in its brief life span. Explosive eruptions from the central crater are often accompanied by lava flows issuing from near the base of the cinder cone. The last eruption was in 1999.

Cosigüina A stratovolcano with a central caldera lake, in western Nicaragua. A single major explosive eruption in 1835 scattered ash throughout Central America and southern Mexico. The ash cloud blotted out the sun over an area 300 kilometers wide. The last eruption was in 1859.

Masaya A caldera with small central stratovolcanoes, in southwestern Nicaragua (see Figure 104). It has erupted many times since 1524 with varied activity, including explosions, lava flows, and lava lakes. The last eruption was in 2003.

PACIFIC OCEAN

Ambrym A stratovolcano within a huge caldera in Vanuatu. Although the record is incomplete, it has been very active since its discovery in 1774, with many ash explosions and lava flows. The last eruption was in 2004.

Anatahan A stratovolcano on a 9-kilometer-long island in the central Mariana Islands. The first historical eruption of Anatahan occurred in 2003, with a large explosive event that created a new crater inside the eastern caldera. The last eruption was in 2004.

Macdonald Seamount A submarine volcano at the southeastern end of the Austral Islands and Seamounts in the South Pacific. Its summit is about 30 meters below sea level. This volcano was discovered to be active in 1967, when noises from it were picked up by distant hydrophones. In 1987 an oceanographic expedition sailing over the seamount witnessed the eruption of large gas bubbles and small fragments of basalt. Seismometers and hydrophones indicated that the eruptions continued into 1989.

Pagan A stratovolcano in the Mariana Islands of the western Pacific Ocean. After four small eruptions between 1909 and 1925, a large explosive eruption in 1981 lofted an ash cloud to 20 kilometers above the island. Smaller eruptions have occurred since, the last in 1993.

Yasour A stratovolcano in Vanuatu. It has been in almost constant mild eruption since its discovery in 1774. Numerous small explosions hurl lava bombs 20 to 200 meters into the air.

PAPUA NEW GUINEA

Karkar A stratovolcano island with a summit caldera, off the north coast of Papua New Guinea. It has erupted about ten times since 1643. The eruption in 1979 took the lives of volcanologists Robin Cooke and Elias Ravian. The last eruption was in 1979.

Lamington A stratovolcano with a lava dome, in eastern Papua New Guinea. This dormant volcano suddenly exploded in 1951. The pyroclastic flows devastated more than 200 square kilometers and killed about 3000 people. A 500-meter-high lava dome grew in the explosion crater from 1951 to 1956. The volcano has been quiet since then.

Manam A 1300-meter-high stratovolcano that forms a circular island 10 kilometers in diameter, off the north coast of Papua New Guinea. Intermittent small explosive eruptions occurred frequently from 1974 to 2003.

Rabaul A group of small volcanoes around the rim of a caldera bay on New Britain Island. Although the vents around Rabaul Harbor have erupted only seven times since 1767, the violent explosive eruption in 1937 formed Vulcan Crater, generated tsunamis, and killed 500 people. An eruption in 1994 killed 8 people; more than 50,000 had evacuated (see Figure 171). The last eruption was in 2004.

PERU

El Misti A classic stratovolcano rising high above the city of Arequipa. Small explosive eruptions, the last in 1784, and gentle steaming characterize its historic activity. It is noted for its beauty rather than its habits.

PHILIPPINES

Mayon A classic stratovolcano cone in the central Philippines. It has erupted explosively more than 40 times since 1616, often producing pyroclastic flows and lava flows (see Figure 103). The region is densely populated, and at least nine eruptions have caused fatalities. The last eruption was in 2003.

Mount Pinatubo An eroded stratovolcano on Luzon Island in the northern Philippines. Until 1991 this volcano had no recorded eruptions; this was suddenly changed by the second largest eruption of the twentieth century (after Katmai). Huge pyroclastic flows and mudflows inundated the surrounding areas. More than 300 people were killed by the eruption, but tens of thousands more would have been killed if the region had not been evacuated. This giant eruption is a classic success story of volcano forecasting (see Chapter 5).

Taal A stratovolcano island within a huge caldera lake, in the central Philippines. It has erupted more than 30 times since 1572, generally with explosions sometimes accompanied by tsunamis in the lake. This killer volcano took many lives in 1716, 1749, 1754, 1874, 1911, and 1965. The last eruption was in 1977.

RUSSIA (KAMCHATKA)

Bezymianny A complex stratovolcano in central Kamchatka. Small explosive eruptions beginning in 1955 climaxed in a gigantic explosion in 1956, forming a mushroom cloud 35 kilometers high (see Figure 59). The thick, hot ash flow deposits formed the Valley of Ten Thousand Smokes of Kamchatka, a feature similar to that created by the Katmai eruption in Alaska in 1912. The last eruption was in 2003.

Karymsky A stratovolcano within an ancient caldera, in southeastern Kamchatka. Its 36 eruptions since 1771 have been mostly explosive. The last eruption was in 2004.

Kliuchevskoi A classic stratovolcano in central Kamchatka. It has erupted explosively more than 80 times since 1697, often producing lava flows. The last eruption was in 2004.

Tolbachik A stratovolcano with a caldera, in central Kamchatka. After about 20 moderate eruptions since 1740, it produced a giant fissure eruption in 1975 that built several large cinder cones and major lava flows totaling nearly 2 cubic kilometers in volume. Forerunning earthquake swarms led to a prediction of the time and place of the 1975 eruption, precise enough that TV camera crews were on location for the outbreak (see Figure 165).

TANZANIA

Ol Doinyo Lengai A stratovolcano—whose name means "mountain of God"—in northern Tanzania (see Figure 147). Although it has erupted more than ten times since 1880, this volcano is more famous to science for its unusual lava flows composed partly of sodium carbonate. The last eruption was in 2004.

UNITED STATES (ALASKA)

Augustine A stratovolcano island with a summit lava dome, in Cook Inlet, southern Alaska. It has erupted explosively seven times since 1812. Danger from major ashfalls or volcanically generated tsunamis worries the offshore oil developers located nearby. The last eruption was in 1986.

Katmai A complex stratovolcano with a summit caldera and lake, on the Alaska Peninsula. The 1912 eruption of Katmai was among the world's largest in historic time. In two days, thick ash deposits covered a huge area, and a pyroclastic flow filled a valley 3 kilometers wide and 20 kilometers long, creating the Valley of Ten Thousand Smokes. The summit caldera, 5 kilometers in diameter, was formed by collapse during the eruption. The total volume of the ash and glowing avalanche was 30 cubic kilometers. Although most of the valley "smokes" (steam vents) are now dead, Mount Trident, a new vent on the west flank of Katmai, became active in 1949 and has exploded and extruded thick lava flows several times. The last eruption of Trident was in 1975.

Pavlof A stratovolcano on the Alaska Peninsula (see Figure 97). It has had more than 40 periods of small to moderate explosive eruptions since 1790. The last eruption was in 1997.

Redoubt A stratovolcano in southwestern Alaska. This peak, visible from the city of Anchorage, has erupted only a few times in recorded history. The 1989–1990 explosive eruptions generated mudflows that threatened oil storage facilities on the north shore of Cook Inlet. The ash cloud almost caused the crash of a large airliner by clogging its jet engines.

Shishaldin A classic stratovolcano cone in the eastern Aleutian Islands. It has a record of about 30 small to moderate explosive eruptions since 1775. Its sharp, snow-covered peak often has a plume of volcanic gases. The last eruption was in 1999.

Mount Spurr A stratovolcano with a large lava dome in southwestern Alaska. Explosive eruptions in 1953 and 1992 deposited small amounts of ash on Anchorage, 130 kilometers east of the volcano. The last eruption was in 1992.

UNITED STATES (CASCADE RANGE)

Lassen Peak A stratovolcano with a summit lava dome, in northern California (see Figure 100). It had a series of explosive eruptions in 1914–1915 that culminated in avalanches, a pyroclastic flow, and mudflows in May 1915. Small explosions continued until 1917.

Mount Rainier A massive, glacier-covered stratovolcano near Seattle, Washington (see Figure 94). It has been dormant since about 1850, when a small explosive eruption apparently occurred. Small steam vents issue from the edges of the snow- and ice-filled summit crater.

Mount St. Helens A snow-covered stratovolcano in southern Washington State, known for its beauty and serenity before its gigantic explosive eruption on May 18, 1980. A directed blast leveled 500 square kilometers of forest, and a major debris avalanche filled a valley for 25 kilometers. The 2950-meter-high summit was lowered by 400 meters, forming a deep horseshoe crater facing north. About 57 people were killed, including David Johnston, a volcanologist. The last eruption, a steam explosion, was in 2004 (see Chapter 4).

UNITED STATES (HAWAII)

Kilauea A shield volcano with a summit caldera, on the island of Hawaii. Famous for its active lava lake during the 1800s and early 1900s, it has also erupted extensive lava flows more than 50 times from both the summit and rift zones. The last eruption was in 2005 and continuing (see Chapter 7 for additional information on Kilauea's Pu'u O'o eruption, which began in 1983).

Mauna Loa A massive shield volcano with a summit caldera, on the island of Hawaii (see Figure 110). It has erupted large lava flows from both the summit and rift zones 38 times since 1832, producing a total of nearly 4 cubic kilometers of basalt. The last eruption was in 1984.

APPENDIX B

Volcanoes on the Internet

Many institutions and individuals have created excellent Web sites that provide information about volcanoes, new eruption data, great photographs, video clips, and even profiles of volcanologists.

In this fourth edition of *Volcanoes* we have listed many Web sites in the "Links" section of each chapter. In addition, we have provided a number of illustrations on the W. H. Freeman Web site that supplements this edition (www.whfreeman.com/volcanoes4e). To avoid information overload, we have limited this appendix to the ten best general Web sites about volcanoes. Remember that Web sites come and go, URLs change, and some sites are updated more frequently than others. Log on and explore volcanoes on the Internet. Getting there is half the fun.

TOP TEN (AS OF 2004, IN ALPHABETICAL ORDER)

Archive of Volcano Photos
http://www.doubledeckerpress.com/archive.htm
Arranged by country, with changing "volcano-photo-of-the-week."

Cascade Volcano Observatory's Big List
http://vulcan.wr.usgs.gov/Servers/earth_servers.html#A
Any and all useful volcano or earth science–related links, both foreign and domestic, government agencies and universities, and so forth . . . if something is missing, please let them know.

Geology.About.Com Volcano Science
http://geology.about.com/od/volcanology/
Twenty-three carefully selected volcano sites.

How Volcanoes Work (California State University, San Diego)
http://www.geology.sdsu.edu/how_volcanoes_work/
This NASA-sponsored Web site is an excellent educational resource that describes the science behind volcanoes and volcanic processes.

Michigan Technological University Volcanoes Page
http://www.doubledeckerpress.com/archive.htm
For serious students, excellent links, frequently updated, good visuals.

Open Directory Volcano Links
  http://dmoz.org/Science/Earth_Sciences/Geology/Volcanoes/
One hundred volcano sites in several categories.

Smithsonian Global Volcanism Program
  http://www.volcano.si.edu/index.cfm
Very comprehensive, worldwide data on 1500 volcanoes that have known and possible eruptions during the past 10,000 years.

Smithsonian Institution & USGS Weekly Volcanic Activity Report
  http://www.volcano.si.edu/gvp/reports/usgs/index.cfm
Lists and describes both new and ongoing activity.

U.S. Geological Survey Volcano Hazards Program
  http://volcanoes.usgs.gov/
Links to volcano data and to the USGS volcano observatories.

Volcano World (University of North Dakota)
  http://volcano.und.nodak.edu/
All ages, fine visuals, great links, frequently updated.

INTERNET SEARCH ENGINES
If these sources don't feature the volcano or offer the information you are looking for, try the conventional search engines such as All the Web, AltaVista, Excite, Gigablast, Google, Hotbot, Infoseek, Lycos, Magellan, OpenText, Teoma, WiseNut or Yahoo, and type in volcano, volcanoes, or the name of the specific volcano you are trying to find.

APPENDIX C

Metric–English Conversion Table

	Length
1 centimeter	0.3937 inch
1 inch	2.5400 centimeters
1 meter	3.2808 feet
1 foot	0.3048 meter
1 meter	1.0936 yards
1 yard	0.9144 meter
1 kilometer	0.6214 mile
1 kilometer	3281 feet
1 mile	1.6093 kilometers

	Area
1 square centimeter	0.1550 square inch
1 square inch	6.452 square centimeters
1 square meter	10.764 square feet
1 square meter	1.1960 square yards
1 square foot	0.0929 square meter
1 square kilometer	0.3861 square mile
1 square mile	2.590 square kilometers

	Volume
1 cubic centimeter	0.0610 cubic inch
1 cubic inch	16.3872 cubic centimeters
1 cubic meter	35.314 cubic feet
1 cubic foot	0.02832 cubic meter
1 cubic meter	1.3079 cubic yards
1 cubic yard	0.7646 cubic meter
1 cubic kilometer	0.2399 cubic mile

Mass

1 gram	0.03527 ounce
1 ounce	28.3495 grams
1 kilogram	2.20462 pounds
1 pound	0.45359 kilogram

Density

1 gram/cubic centimeter	62.4280 pounds/cubic feet

Pressure

1 kilogram/square centimeter	0.96784 atmosphere
1 kilogram/square centimeter	0.98067 bar
1 kilogram/square centimeter	14.2233 pounds/square inch
1 bar	0.98692 atmosphere
1 atmosphere	1.0332 kilogram/square centimeter

Temperature

$(\text{Celsius} \times \frac{9}{5}) + 32$	Fahrenheit
$(\text{Fahrenheit} - 32) \times \frac{5}{8}$	Celsius

Energy

1 erg	2.39006×10^{-8} gram calorie
1 joule	10^7 ergs
Explosion equivalent to 1000 tons of TNT	4×10^{19} ergs

Power

1 watt	10^7 ergs/second
1 watt	0.001341 horsepower

Glossary

a'a A type of lava flow with a rough, fragmental surface.

active volcano A volcano that is erupting or has erupted in recorded history.

airfall deposit Pyroclastic fragments that have fallen from an eruption cloud.

andesite A volcanic rock type intermediate in composition between dacite and basalt.

angle of repose The steepest slope at which loose material will come to rest without slumping.

ash cloud An explosive eruption cloud containing large quantities of volcanic ash.

ash flow A pyroclastic flow of mainly ash-sized (less than 2 millimeters in diameter) material.

ash, volcanic Fine pyroclastic material, down to dust size, that is formed by explosive volcanic eruptions.

asthenosphere A zone of low strength in the Earth's upper mantle. See also *low-velocity layer*.

avalanche A large amount of snow, ice, soil, and/or rock falling or sliding at high speed.

barograph An instrument that makes a continuous record of changes in atmospheric pressure.

basalt A dark, heavy lava composed of about 50 percent silica, 15 percent alumina, and significant amounts of iron, calcium, and magnesium.

base surge A ground-hugging, expanding ring of gas and debris at the base of an explosion column.

block, volcanic A solid fragment thrown out in an explosive eruption, ranging in size from about 6 centimeters to several meters in diameter.

bomb, volcanic A still-viscous lava lump thrown out in an explosive eruption, which takes on its rounded shape while in flight.

caldera A gigantic basin with steep walls at the summit of a volcano; larger than a crater, and usually formed by collapse.

cinder cone A steep conical hill formed by the accumulation of cinders and other loose material expelled from a volcanic vent by escaping gases.

cinder, volcanic A pyroclastic fragment about 1 centimeter in diameter.

composite cone See *stratovolcano*.

conduit A pipe or crack through which magma moves.

continental crust The thick layer of granitic rocks beneath the continents.

continental drift The theory that horizontal movements of the Earth's surface cause slow relative movements of continents toward or away from one another.

convergent margin The edges of tectonic plates that are moving together.

crater A bowl- or funnel-shaped depression, generally in the top of a volcanic cone; often the major vent of volcanic products.

crystalline rock A hard rock composed of interlocking crystals; often of igneous origin.

dacite A volcanic rock type intermediate in composition between rhyolite and andesite.

debris avalanche An avalanche composed of a mixture of rock, soil, ice, snow, trees, and so forth.

debris flow A rapid flow of debris of various kinds; for example, a high-density mudflow. Debris avalanches become debris flows after mixing with water.

debris flow, cohesive A debris flow with a high clay content that behaves like wet concrete.

debris flow, noncohesive A debris flow with a low clay content that deposits its coarse and dense material as it loses velocity.

dike A sheetlike body of intrusive igneous rock that cuts across the layering of the host rock.

divergent margin The edges of tectonic plates that are moving apart.

dormant volcano A volcano that is not presently erupting but is considered likely to do so in the future.

dust, volcanic The finer particles of volcanic ash.

earthquake swarm A sequence of closely spaced earthquakes during which the energy release is approximately constant (as opposed to a sequence consisting of a large quake followed by diminishing aftershocks).

earthquake wave A general term for a vibrational wave produced by an earthquake.

effusive eruption An eruption consisting mainly of lava flows (as opposed to an explosive eruption).

eruption cloud A gaseous cloud of volcanic ash and other pyroclastic material that forms by volcanic explosion.

eruption column The trunk or lower portion of an eruption cloud that connects the cloud to the volcanic vent.

extinct volcano A volcano that is not erupting and is not expected to do so in the future; a dead volcano.

fault A fracture in the Earth's crust along which there has been movement.

feldspar A light-colored mineral composed largely of silicon, oxygen, and aluminum.

flank eruption An eruption from the side of a volcano (in contrast to a summit eruption).

fume A gaseous cloud without volcanic ash.

gabbro A coarse-grained igneous rock of basaltic composition.

geophysics The physical and mechanical aspects of geology (in contrast to geochemistry).

geothermal energy Energy derived from the internal heat of the Earth.

geothermal power Power generated by using the heat energy of the Earth.

granite A coarse-grained igneous rock composed mostly of quartz and feldspar.

hazard A potentially dangerous phenomenon, such as a pyroclastic flow or mudflow. See also *risk*.

heat transfer Movement of heat from one place to another.

hot-spot volcanoes Volcanoes related to a persistent heat source in the Earth's mantle.

hydrothermal reservoir An underground zone of porous rock containing hot water.

hyperconcentrated flow A transitional state between a mudflow and seasonal muddy-water stream flow.

igneous rock Rock formed from the cooling and solidification of magma or lava.

ignimbrite A rock formed by the consolidation of pyroclastic flow deposits.

intrusion An igneous rock body formed when molten igneous rock forces its way into surrounding host rocks and then cools; also, the process of forming such an igneous rock body.

island arc A curving chain of volcanic islands formed at a convergent plate boundary.

Kimberlite pipe A vertical pipe-shaped intrusion of unusual igneous rocks that often contain diamonds.

lahar An Indonesian name for a volcanic mudflow.

lapilli Pyroclastic fragments about 1 centimeter in diameter. *Lapilli* is an Italian word meaning "little stones."

lateral blast A hot, low-density mixture of rock debris, ash, and gases, generated by a volcanic explosion, that moves at high speed along the ground surface.

lava Magma or molten rock that has reached the Earth's surface; also, the rock formed by its cooling and solidification.

lava dome A steep-sided, rounded extrusion of highly viscous lava, squeezed out from a volcanic vent.

lava lake A lake of molten lava in a volcanic crater or depression; refers to the solidified and partly solidified stages as well as to an active lava lake.

lava tube A tunnel formed when the surface of a lava flow cools and solidifies and the still-molten interior lava flows through; also, a hollow tube that remains when the interior lava drains away.

linear vent A vent formed by a long fissure reaching the Earth's surface (in contrast to a single crater).

lithosphere The zone in the crust and upper mantle of the Earth that exhibits strength.

low-velocity layer The zone in the upper mantle, about 60 to 250 kilometers in depth, in which seismic wave velocities are lower than in the overlying layers.

maar A wide, low-relief crater formed by volcanic explosion, generally filled with water.

magma Molten rock material containing dissolved gases that forms igneous rock upon cooling; magma that reaches the Earth's surface is called *lava*.

magma chamber An underground reservoir in the Earth's crust filled with magma, from which volcanic materials are derived.

magmatic fluids Volcanic gases, especially water and carbon dioxide, dissolved in magma.

magnetic field A region in which magnetic forces exist.

mantle The zone of the Earth below the crust and above the core (to a depth of 3480 kilometers).

microearthquake An earthquake that is not felt but is detectable by a seismograph.

mudflow A flowing mixture of water-saturated mud and debris that moves downslope under the force of gravity.

neck A vertical, pipelike intrusion that represents a former volcanic vent; usually used to describe an erosional remnant.

normal fault An inclined fault in which the upper block moves relatively downward.

nuée ardente A fast-moving, dense "glowing cloud" of hot volcanic ash and gas erupted from a volcano.

obsidian A black or dark-colored volcanic glass, usually composed of rhyolite.

oceanic crust The Earth's crust where it underlies the oceans, without the granitic layer that forms continents.

olivine An olive-green mineral composed of iron, magnesium, silicon, and oxygen.

ore The naturally occurring material from which a mineral, or minerals, of commercial value can be extracted.

pahoehoe A type of lava flow with a smooth, billowy, or undulating surface.

partial crystallization The stage of cooling of magma when it is partly solid crystals and partly liquid rock.

partial melt The stage of melting of rock when it is partly liquid rock and partly solid crystals.

pillow lava Interconnected, sacklike bodies of lava formed under water.

plate tectonics The theory that the Earth's crust is broken into about 12 large plates that slowly move about the surface.

plume A rising column of magma from deep in the mantle responsible for hot-spot volcanoes; sometimes refers to an ash or fume cloud.

pluton A large igneous intrusion formed at depth in the crust.

precipitate A solid forming from a solution.

pumice A form of volcanic glass so filled with gas bubble holes that it resembles a sponge and is very light.

pumice flow A pyroclastic flow of mainly pumice fragments.

pyroclastic deposit The material deposited by a pyroclastic flow or pyroclastic fall. The word *pyroclastic* should be used as an adjective.

pyroclastic fall The fallout material from an eruption cloud, also called tephra.

pyroclastic flow A fluidized mass of hot, dry rock fragments mixed with hot gases that moves away from a volcano at high speeds.

pyroclastic surge A variety of pyroclastic flow caused by a base surge.

quartz An important rock-forming mineral composed of silicon and oxygen (SiO_2).

rhyolite A fine-grained volcanic rock with the same composition as granite; although rhyolite and granite have the same composition, they differ in texture.

ridge, oceanic A major submarine mountain range.

rift system The oceanic ridges, totaling over 60,000 kilometers in length, formed where plates are separating and new crust is being created; also, their subaerial equivalents such as the East Africa Rift.

Ring of Fire The regions of mountain-building earthquakes and volcanoes that surround the Pacific Ocean.

risk The danger to life or property posed by a hazard.

seafloor spreading The mechanism by which new seafloor crust is created at oceanic ridges and slowly spreads away on the separating plates.

seamount An isolated tall mountain on the seafloor, generally volcanic.

sedimentary mud Loose fine-grained sediment with enough water to form soft mud.

seismic wave See *earthquake wave.*

seismograph An instrument that records the motions of the Earth's surface caused by seismic waves.

seismology The study of earthquakes, seismic waves, and the structure of the interior of the Earth.

shearing The motion of two surfaces sliding past each other.

shield volcano A gently sloping volcano in the shape of a flattened dome, built by flows of very fluid basaltic lava.

silica A chemical combination of silicon and oxygen.

silicate mineral A mineral largely composed of silicon and oxygen.

slump A slow or intermittently moving landslide or submarine slide.

stock A large igneous intrusion roughly circular in a horizontal plane.

stratovolcano A steep volcanic cone built by both lava flows and pyroclastic eruptions.

strike-slip fault A nearly vertical fault with side-slipping displacement.

subduction zone The zone of convergence of two tectonic plates, one of which usually overrides the other.

tephra A general term for all airfall pyroclastic material from a volcano; sometimes used as a synonym for all types of pyroclastic material.

thermal gradient The rate of change of temperature with distance or depth.

thermocouple A pair of twisted bare wires of different metals that generates voltage when heated; a thermoelectric device used to measure molten lava temperature.

thrust fault A gently inclined fault whose upper side moves relatively upward.

tidal wave See *tsunami*.

transform fault A strike-slip fault connecting the offsets of mid-ocean ridges.

tsunami A great sea wave produced by a submarine earthquake or volcanic eruption.

tuff A compacted pyroclastic deposit that generally contains inclusions of sedimentary material.

tuff ring The rim of explosion debris (tuff) that surrounds a wide, shallow crater.

vein A mineral deposit precipitated in a rock fracture.

vent An opening at the Earth's surface through which volcanic materials issue forth.

viscosity A measure of resistance to flow in a liquid; water has a low viscosity, while honey has a high viscosity.

volcanic front The line of volcanoes closest to the oceanic trench in an island arc like Japan.

wave-cut terrace A level surface formed by wave erosion of coastal rocks; may appear above sea level if uplifted.

welded tuff A hard pyroclastic rock compacted by internal heat and the pressure of overlying material.

Index

Note: Page numbers followed by f indicate illustrations; those followed by t indicate tables.